Do Brilliantly

AS Biology/Human Biology

Margaret Baker
Alan Morris
Series Editor: Jayne de Courcy

Published by HarperCollins*Publishers* Limited
77–85 Fulham Palace Road
London W6 8JB

www.**Collins**Education.com
On-line support for schools and colleges

First published 2001

ISBN 0 00 710703 X

British Library Cataloguing in Publication Data
A catalogue record for this book is available from the British Library

Edited by Jane Bryant
Production by Kathryn Botterill
Cover design by Susi Martin-Taylor
Book design by Gecko Limited
Printed and bound by Scotprint, Haddington

Acknowledgements
The Authors and Publishers are grateful to the following for permission to reproduce copyright material:
AEB, a division of AQA (pp. 12, 13, 14, 22, 28, 45, 48, 53, 54, 55, 56, 61, 68, 75, 76). Answers to questions taken from past examination papers are entirely the responsibility of the author and have neither been provided nor approved by the AEB.
AQA (pp. 7, 10, 77). Answers to questions taken from past examination papers are entirely the responsibility of the author and have neither been provided nor approved by the AQA.
Northern Examinations and Assessment Board, a division of AQA (pp. 20, 23, 29, 30, 39, 46, 62, 63, 69, 70). The author is responsible for the suggested answers and the commentaries on the past questions from the Northern Examinations and Assessment Board. They may not constitute the only answers.
OCR (pp. 19, 31, 35, 37, 47). Answers to questions taken from past examination papers are entirely the responsibility of the author and have neither been provided nor approved by the OCR.

Illustrations
Cartoon Artwork – Roger Penwill
DTP Artwork – Geoff Ward

Photographs
CNRI/Science Photo Library (p.26); Ken Eward/Science Photo Library (pp.33, 34); T.J. O'Donnell, Custom Medical Stock Photo/Science Photo Library (p.34); Geoff Tompkinson/Science Photo Library (p.44).

Every effort has been made to contact the holders of copyright material, but if any have been inadvertently overlooked, the Publishers will be pleased to make the necessary arrangements at the first opportunity.

Contents

How this book will help you

by Margaret Baker and Alan Morris

Exam practice – how to answer questions better

This book will help you improve your performance in your AS Biology or Human Biology exam. It contains lots of **questions on the core topics of the new AS specifications**.

In exams, students often fail to gain the grades they are capable of, even when they have worked hard throughout their course and have a sound grasp of biological facts. It is often not a lack of knowledge that leads to disappointing results, but poor examination technique. **This book will help you with your examination technique so that you can make the most of your knowledge and score high marks in your exam**.

Each chapter in this book is broken down into four separate elements, aimed at giving you as much guidance and practice as possible:

1 Exam question, Student's answer and 'How to score full marks'

The questions and students' answers that we have chosen to start each chapter are typical ones. They show a number of mistakes frequently made by candidates under exam conditions.

The 'How to score full marks' section explains precisely where the student lost marks, e.g. not following instructions in the question, not putting in enough detail, etc. In each case, **we show you how to gain the extra marks so that when you meet these sorts of questions in your exam, you will know exactly how to answer them.**

2 'Don't make these mistakes'

This section highlights the most common mistakes we see every year in exam papers. These include poor use and understanding of biological terminology and not interpreting the questions correctly. **When you are into your last minute revision, you can quickly read through all of these sections and make doubly sure that you avoid these mistakes in your exam**.

3 'Key points to remember'

These key facts pages have been designed to be as concise as possible. Here you will find much of what you need to know in order to answer questions on the core topics in your AS specification. **They are the most important points that you need to cover when revising a particular topic**. You will also need to use your own notes and/or your textbook.

4 Questions to try, Answers and Examiner's comments

Each chapter ends with a number of exam questions for you to answer. Don't cheat. Sit down and answer the questions as if you were in an exam. Try to put into practice all that you have learnt from the previous sections in the chapter. We've included some exam hints before each question which should help you get the correct answers. Check your answers through and then look at the answers given at the back of the book. These are full mark answers.

In our 'Examiner's comments' we highlight anything tricky about the question which may have meant you did not get the correct answer. By reading through these sections, you can avoid making mistakes in your actual exam.

The topics covered by your AS specification

This book is divided into nine chapters **covering all the core Biology and Human Biology topics set by all the exam boards**, and a couple of the most common 'extra topics' covered in AS specifications. The chart below lets you see at a glance which topics your particular specification requires you to study at AS level and so which topics will be in questions on your exam paper.

TOPICS COVERED (Chapters in this book)	EXAMINATION BOARDS			
	AQA Specification A	AQA Specification B	OCR	EDEXCEL
Enzymes and their properties	✓	✓	✓	✓
Getting into and out of cells	✓	✓	✓	✓
Protein synthesis and genetic engineering	✓	✓	✓	✓
Biological molecules	✓	✓	✓	✓
Mitosis, meiosis and the cell cycle	✓	✓		
Nucleic acids	✓	✓	✓	✓
Cells, organelles and sizes	✓	✓	✓	✓
Ventilation and gas exchange	✓			
The heart and circulation	✓			

Tips on revising

Plan of action

- Make sure you give yourself **plenty of time**, but not too much that it becomes ineffective.
- Work out **a revision plan**. Start with the topics that you find the most difficult or that you know least about.
- Make sure your plan is **realistic**, and allows some extra time in case a topic takes longer than you had imagined.
- Stick to your plan!

Quality revision time

Few people can work effectively for more than an hour at a time. After this they lose concentration. **Revise in short, concentrated bursts** then have a break, a walk in the garden, a cup of tea!

Learn the facts and practise questions

Start your revision by making sure you have **a really good grasp of the basic principles**. You'll then find it much easier to learn the necessary **detail**. Once you've revised the facts, give yourself **lots of practice** at answering questions to make sure that you can turn your knowledge into high exam marks.

Read the question carefully

Examiners try to make the wording of the questions as straightforward as possible, but in the stress of an examination it is all too easy to misinterpret a question. **Read every word in every sentence very carefully**. The paper is yours and no examiner will be cross if you underline key words to help you understand exactly what you have to do.

Understand the information

Not all AS questions are based on recall of knowledge. Some are based on unfamiliar material. It is important that you do not panic and are able to use **the principles that you have learnt** to answer these types of questions.

Make sure **you always identify the area of the specification that the question is about**, e.g. 'enzymes' – then you can start to think in terms of proteins, substrate, collisions, etc.

Look at information given to you in a graph or a table very carefully and make sure you give yourself time to **read the information** on the axes or the headings of the table. Make sure you understand what this information means **BEFORE** you read the questions.

Look at the marks

Each question or part of a question is given a number of marks. You must **make sure that your answer gives enough information to get all the marks**. For example, if the question is worth 4 marks then you must give 4 points to gain those marks.

Know the meaning of common instructions

Here is a list of some of the common words used in AS Biology and Human Biology exam questions. **Make sure you are familiar with them and that you know what an examiner expects you to do**. Check your exam papers for any other action words and be sure you know what they mean too.

- **Describe**
Replace this word with the word '**what**'. If this term is used in relation to a graph or a table, then you are being asked to recognise a simple trend or pattern within the data and write what it is. You must use the information on the axes or table headings as reference points. 'It goes up and then down' is not enough; what goes up and by how much would be expected at this level.
In longer questions 'describe' means you need to give a step-by-step account of what is happening.

- **Explain**
Replace this word with the word '**why**'. This means you must give a biological reason for the pattern you are given in the question. It does not mean the same as describe and, although you may need to describe the pattern before you can explain it, no marks will be given if you only describe what is happening.

- **Suggest**
This is often used when you are not expected to know the answer, but should have enough biological knowledge to put forward a sensible idea.

- **Name**
This means exactly what it says and requires no more than a one-word answer. You do not have to repeat the question or put your answer into a sentence, e.g. 'The name of the organelle is a mitochondrion.' This only wastes your time.

Exam question and student's answer

Read the following passage.

> Starch is an important storage carbohydrate in most plants and is found as insoluble granules in the cytoplasm. It is a polymer of glucose.
>
> Corn starch is a very cheap substance and can be converted to fructose using various enzymes obtained from microorganisms. The starch suspension is heated to a temperature of 105°C and an enzyme called alpha-amylase obtained from bacteria is added. It is a thermostable enzyme and it begins to hydrolyse the starch. The temperature is then reduced to 90°C and hydrolysis continues for 1–2 hours. The bonds within the polysaccharide molecule are hydrolysed and the long chains are broken into smaller chains called dextrins.
>
> The next step involves the conversion of dextrins to glucose and this is carried out by a fungal enzyme called amyloglucosidase. The substrate is adjusted to conditions producing the maximum rate of reaction, a temperature of 60°C and a pH of 4.5, before the enzyme is added. Amyloglucosidase removes glucose, one molecule at a time, from the end of the dextrin molecule.
>
> After concentration, the resulting glucose syrup can be converted into fructose. This process involves glucose isomerase produced by bacteria.

Describe and explain why the rate of reaction of amyloglucosidase would vary with temperature.

[Total 6 marks]

As the temperature increases so does the rate of reaction. Increasing the temperature means more energy is available and this makes the ✓ enzyme molecules and the substrate molecules move about more. This will increase the number of collisions ✓ and so more end products will be formed. At high temperatures the enzyme is denatured. ✓

3/6

How to score full marks

What and why?

This question asks both for a description (what is happening) and an explanation (why it is happening). In many cases you can't do one without the other. This answer does describe the effect of increasing the temperature on the rate of reaction in its first sentence, but it does so incompletely, for there is no mention of the **effect at high temperatures**. Both ideas are needed for a mark.

The principle

The principle of thermal energy (heat) increasing the kinetic energy (movement) of molecules is given in the answer. **Marks are often awarded for the principle alone** – if you don't use the correct technical terms, as long as you make 'the principle' clear, you will still get a mark. However, **an extra mark may be given for the use of those terms**. In this case the term **'kinetic energy'** would have been given a mark.

Don't leave out steps

This answer leads logically onto the idea of increased collisions, but doesn't link the fact that **this must be between substrate and enzyme molecules for a reaction to occur**. The formation of an enzyme–substrate complex should be mentioned for a mark.

'A'-level quality

The answer mentions denaturation but offers no explanation in terms of the change in shape of the tertiary structure. **You can only get one mark for one word.** Denaturation is a process, but an **explanation** of the process would be awarded more marks.

Don't make these mistakes ...

If you are asked to 'describe', do so in full using the data given. The term 'describe' is asking you to look for trends or patterns, so write about all the trends you can find.

Only answer the question that is set. There is evidence from the data that different enzymes are affected by temperature in different ways, but don't think you have to use this right away – there may be another question which asks for this idea.

Don't read the question too quickly and miss important information. Take your time when reading the question paper and don't panic – plenty of time is allowed for reading and answering the questions. The information given is intended to help.

Don't use terms like 'denaturation' without explaining their meaning. You need to give the 'full story', especially if there are several marks allocated to this section of the question.

Enzymes are not 'killed' by changes of temperature. As biological molecules they don't have a life force!

Be aware that quantitative information (numbers in the passage or a table or a graph) ought to be used in your answer. For example, **quote the range of values you are describing**.

Key points to remember

Collision theory

Enzyme molecules and substrate molecules in a solution are moving around. They must collide with each other before they will react. **The more collisions that take place, the greater the rate of reaction**.

Activation energy

Energy has to be provided to start any reaction. This is to overcome the natural stability of the molecules involved, and is known as the **activation energy**. **Enzymes lower the activation energy** – they reduce the amount of energy needed.

The lock and key model:

- The active site of an enzyme molecule acts rather like a lock. It has a particular shape into which the substrate molecule will fit.

The induced fit model:

- The active site of an enzyme molecule has a shape which is almost complementary in shape to the substrate molecule. As the substrate enters, the active site moves to fold around the substrate.

Whichever model you consider, **if the shape of the active site is changed, the substrate won't fit** and the enzyme won't function.

At low temperatures ...

- increasing the temperature provides more heat energy
- both the enzyme and substrate move faster
- the chances of the enzyme and substrate colliding increase
- the more collisions, the faster the rate of reaction.

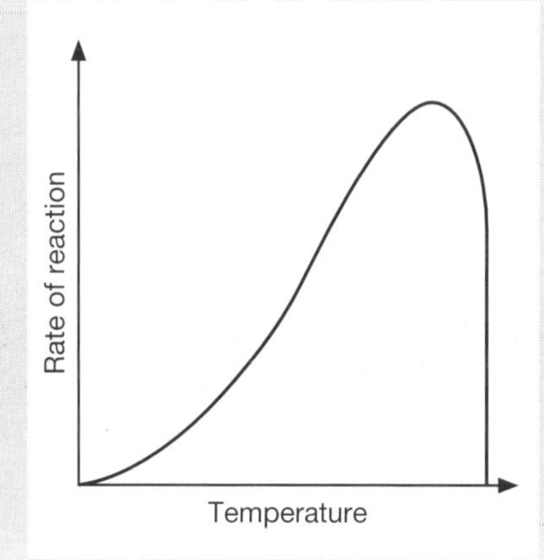

At high temperatures ...

- heat energy makes the bonds holding the tertiary structure of the protein vibrate
- the weakest bonds – the 'hydrogen bonds' – break
- the protein shape changes – it denatures
- the shape of the active site changes
- the substrate no longer fits and the enzyme stops functioning.

Substrate concentration

A: The more substrate molecules there are, the greater the chance of collision with an enzyme molecule and the faster the rate of reaction.

B: All the enzyme active sites are occupied, and only by adding more enzyme molecules will the rate of the reaction increase.

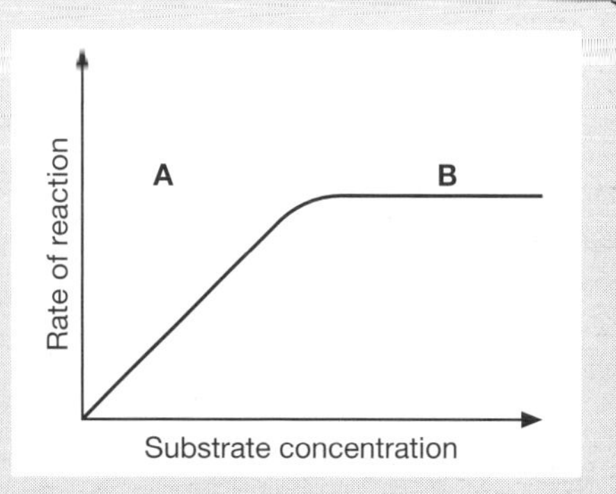

Specificity

- Only the right substrate will fit the active site of the enzyme.
- Different substrate molecules have different shapes and so they will not fit the active site.

pH

- A change of pH may affect the ability of an amino acid to form ions.
- This may lead to breakage of bonds that hold the tertiary structure of the enzyme.
- The shape of the enzyme and the active site changes.
- The substrate no longer fits and the enzyme stops functioning.

Questions to try

Examiner's hints

- Some questions don't ask you to use the information given in the passage or on a graph, but it's always an advantage to do so.
- Some questions specifically ask you to use the information given – and unless you do so, you will not be awarded full marks.
- The exam paper is *yours* – there is no reason why you can't highlight areas of a graph to make explanations clearer.

Q1

Enzymes are catalysts which catalyse specific reactions by lowering their activation energy. The lock and key and the induced fit models have been used to explain the way in which enzymes work.

(a) Explain what is meant by activation energy.

....

....

[1 mark]

(b) (i) Describe how the lock and key model can be used to explain how an enzyme breaks down a substrate molecule.

....

....

....

....

[3 marks]

(ii) Describe how the induced fit model differs from the lock and key model of enzyme action.

....

....

[1 mark]

Catalase is an enzyme found in many cells. It catalyses the breakdown of hydrogen peroxide as shown in the equation.

$$\text{hydrogen peroxide} \xrightarrow{\text{catalase}} \text{water + oxygen}$$

Cylinders of potato were cut with a cork borer. The cylinders were then sliced into discs of the same thickness and put into a small beaker containing 50 cm^3 of hydrogen peroxide. The mass of the beaker and its contents was recorded over a period of 15 minutes. The results are shown in the graph.

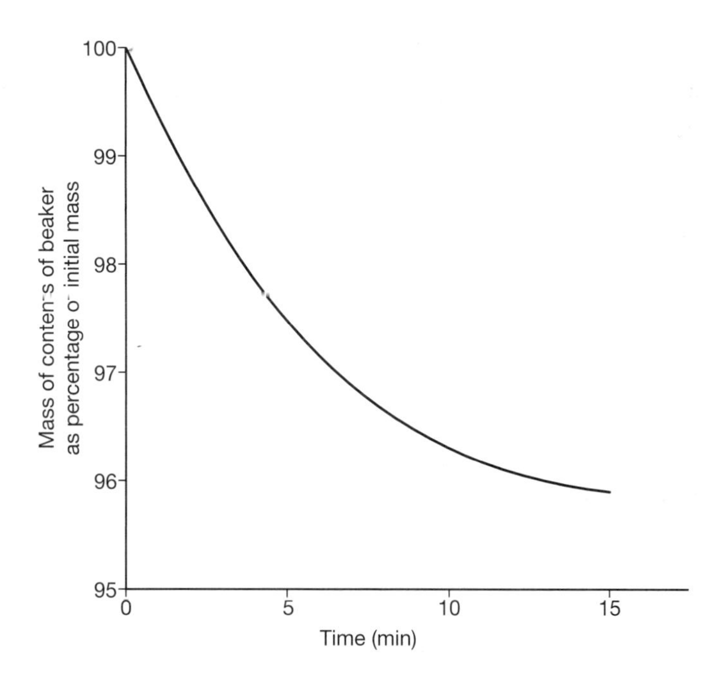

(c) Explain why the mass of the contents of the beaker fell as the reaction progressed.

....

....

[1 mark]

(d) Explain, in terms of collisions between enzyme and substrate molecules, why the rate of the reaction changed over the period of time shown on the graph.

....

....

....

[2 marks]

[Total 8 marks]

Examiner's hints

- If the question introduces terms you don't recognise – like turnover number – take time to read through the exact definition to be sure you understand what it means.
- Think simple. An independent variable is the one that you change (it is always plotted on the *x* axis – the bottom of the graph). A dependent variable is one that you measure (it is always plotted on the *y* axis – the side of the graph).
- If you are asked to give a reason for using a control, **never** use the phrase 'because it is a fair test' at 'A' level.

Q2

The *turnover number* of an enzyme is defined as the number of substrate molecules converted to product by one molecule of enzyme in one minute. In an experiment carried out at 20°C the turnover number for an enzyme was found to be 2500 at the start of the experiment but dropped to 1000 after 5 minutes.

(a) (i) Suggest why the turnover number decreased after 5 minutes.

....

....

....

[2 marks]

(ii) How would you expect the turnover number to differ from 2500 at the start of an identical experiment but carried out at 30°C? Explain your answer.

....

....

....

[2 marks]

(b) Explain why it would be important to have a control in the experiment at 20°C and at 30°C.

....

....

[1 mark]

[Total 5 marks]

Examiner's hints

- If the stem of the question is separated from a number of sub-sections, as in this case, be sure you read the stem again with each sub-section. It is too easy to read the sub-section and miss exactly what you have been asked to do.
- The term 'suggest' in the stem of a question usually means that this material is not required by the syllabus – but you should be able to come up with a reasoned explanation by using the principles you do know. There will often be a number of acceptable answers.

Q3

The diagram shows a metabolic pathway in which substrate A is converted, with the aid of enzymes, to the end-product D.

$$A \xrightarrow{\text{Enzyme 1}} B \xrightarrow{\text{Enzyme 2}} C \xrightarrow{\text{Enzyme 3}} D$$

(a) Giving an explanation for your answers, suggest what would happen to the rate of production of the end-product D if:

(i) The concentration of substrate A were reduced.

....

....

[1 mark]

(ii) The concentration of enzyme 1 were increased but the concentrations of the other enzymes remained constant.

....

....

[1 mark]

(iii) The temperature rose from 15°C to 25°C.

....

....

[1 mark]

(b) Suggest how molecule D could act as an *end-product inhibitor*.

....

....

....

[2 marks]

[Total 5 marks]

Answers to Questions to try are on pp. 78 – 79.

Exam question and student's answer

(a) An artificial membrane can be made which consists only of a lipid bilayer. The diagram compares the permeability of such an artificial membrane with a biological cell membrane.

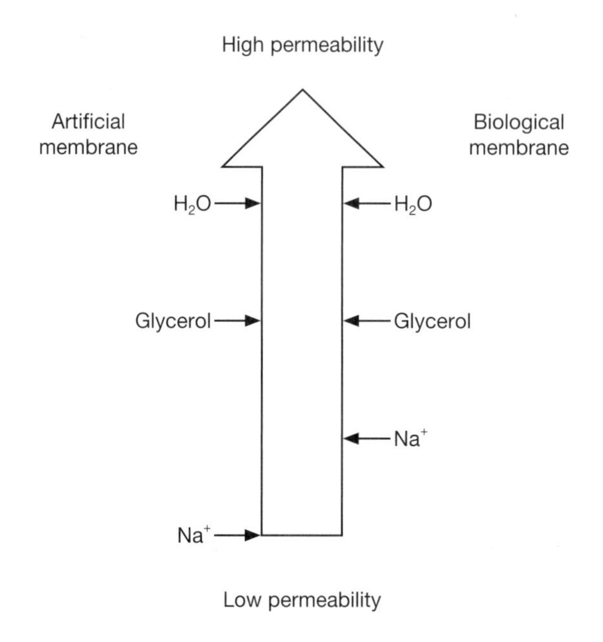

(i) Explain why the permeability of both membranes to glycerol is the same.

Glycerol is a part of a lipid molecule when combined with three fatty acids. So as it is a lipid it moves through both membranes just as easily. ∧

0/1

(ii) Explain why the permeability of the two membranes to sodium ions differs.

Sodium ions move through membranes by active transport. ✓ Biological membranes allow this to happen but artificial membranes do not.

[3 marks]

(b) The diagram shows two chambers separated from each other by a partially permeable membrane and containing the same volume of different concentrations of glucose solution.

A	B
Glucose solution concentration 0.2 mol dm^{-3}	Glucose solution concentration 0.5 mol dm^{-3}

(i) In which of the chambers, A or B, is the water potential higher?

A ✓

1/1

(ii) Give an explanation for your answer.

Chamber A contains the weaker solution. ^

(iii) Explain in terms of water potential why water molecules flow from chamber A to chamber B.

The water potential in chamber A is higher ✓ than in chamber B and as water moves from a higher to a lower ✓ water potential, it moves from A to B.

[5 marks]

[Total 8 marks]

How to score full marks

What is the question about?

Part (a) asks for an understanding of the way membrane structure affects the movement of materials through them.

Is there a clue?

The stem of the question gives the clue that **both membranes are lipid bilayers**.

Remember the principle

In this example, **molecules move through membranes either by diffusion or by active transport. Both involve energy:** diffusion relies on the kinetic energy of the molecules, active transport on ATP generated by the cell. Diffusion does not involve protein carriers, whereas active transport does.

Focus on the right side of the argument

In part (i) the student has recognised that glycerol is part of a lipid. However, the student has missed the point of the question: **the membrane's property to allow free diffusion of these molecules**.

State the obvious

In part (ii) the mechanism, active transport, has been identified but the last sentence adds nothing – it simply amplifies the question. The lack of protein in the artificial membrane and therefore **its inability to assist the movement of sodium ions** was missed.

Never leave a space

When there is a 50/50 choice (as there is in part (b)), even if you have no idea of the answer, guess! If you can't answer part (ii) **you stand a chance of getting one mark by having a go at the first part of the question**. Most questions 'stand alone' and examiners try never to allow candidates to compound errors.

Answer the question

A weaker solution will have more water molecules than a stronger solution but there is no 'explanation' as to why a weaker solution should have a high water potential. **Decreasing the concentration of solute makes the water potential less negative**. Keep this link in your mind.

Higher or less negative?

There are two ways of referring to water potential: **high, or less negative, or low, or more negative, water potential. Water will move from high to low or less negative to more negative**. Both are correct.

Don't make these mistakes ...

Don't answer the wrong question. If the question has two aspects (here the structure of the membrane and the nature of the molecules that move through them) **look carefully to be sure you are answering from the right angle.**

Never restate the question. Sometimes it is difficult to see that this is what you have done. In this case, unless there was a difference in the way the membranes affected the molecules that passed through them, there would not be a difference in their permeability.

Watch the mark allocation carefully. If two marks are available you will need two correct ideas to get them.

Key points to remember

Membrane structure

- Molecules that dissolve in lipids can pass through the bilayer.
- Water-soluble molecules pass through protein pores that span the bilayer.

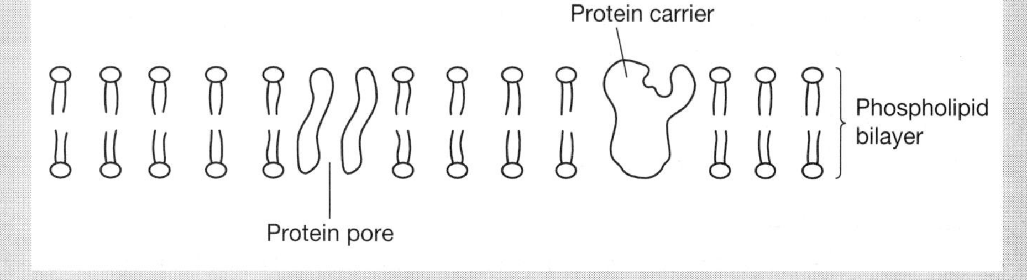

- Proteins act as carriers and have an important role in facilitated diffusion and active transport.

Diffusion through the membrane

- Small or non-polar molecules (e.g. oxygen, carbon dioxide) diffuse freely through the bilayer, between the phospholipid molecules.
- Molecules move from where they are in high concentration to where they are in lower concentration, down a concentration gradient.
- Energy for the movement comes from the kinetic energy of molecules themselves.
- Diffusion does not need an input of energy from the cell in the form of ATP. It is a passive process.
- Diffusion is slow so it is only useful when distances are very small.
- The cell membrane (and therefore the organism) has no control over the pace or direction of movement.

Facilitated diffusion

- Molecules that can't pass directly through the bilayer (e.g. glucose) move by facilitated diffusion.
- Molecules move from high concentration to lower concentration, down a concentration gradient.
- Facilitated diffusion relies on carrier proteins to speed up movement across the membrane.
- The more carrier proteins there are, the faster the possible rate of diffusion.
- These proteins are very specific.
- Different membranes have different carriers.
- The carrier protein can change shape.
- Facilitated diffusion does not need input of energy from the cell in the form of ATP. It is a passive process.
- Proteins in the membrane can speed up diffusion but use the energy of the moving molecules. The organism has no control over the pace or direction of movement.

Protein 'carrier' speeds up movement through the membrane

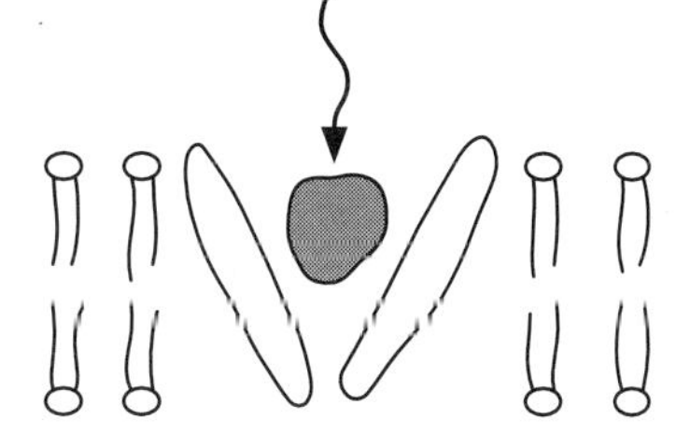

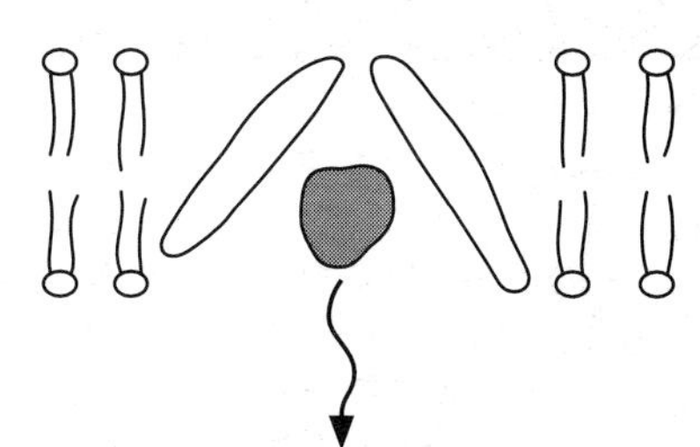

The kinetic energy of the molecule causes the 'gate' to open

Key points to remember

Active transport

- Molecules that cells need, and which are low in concentration in the environment (e.g. nitrate ions), move into the cell by active transport. They can't pass directly through the bilayers.
- Active transport requires a carrier protein.
- These carrier proteins are very specific.
- This process enables a cell to take up substances against a concentration gradient.
- It requires energy, in the form of ATP, which comes from respiration.
- Cells where lots of active transport takes place have many mitochondria to produce ATP.

Water movement

- Water will move from an area of high (less negative) to an area of low (more negative) water potential.
- Water molecules will diffuse across a partially permeable membrane from a solution with a higher water potential to a solution with a lower water potential. As the membrane is partially permeable solutes will not freely diffuse between cells
- This water movement is called osmosis.

Rate of diffusion

Fick's law states that:

$$\text{Rate of diffusion} \propto \frac{\text{Surface area} \times \text{difference in concentration}}{\text{Thickness of exchange surface}}$$

A surface that is adapted for efficient diffusion will:

- have a surface area as large as possible
- maintain a large difference in concentration
- have as thin an exchange surface as possible.
- An increase in temperature will increase the rate of diffusion, due to an increase in kinetic energy of the molecules.

Active transport has a saturation effect

- As the external concentration increases and the concentration difference gets greater, the rate of simple diffusion increases. This pattern is also true of facilitated diffusion and active transport. Eventually the proteins are transporting molecules at their fastest rate (maximum turnover rate) and even if the concentration increases can't increase the rate of movement further.

Water potential

- As water molecules move randomly within a cell, some of them will hit the plasma membrane. This collision will create a pressure. This pressure is called the water potential, ψ, and is measured in kilopascals (kPa).
- The more free water molecules present, the greater the water potential.
- Pure or distilled water has the highest water potential, which is given the value zero.

The effect of soluble molecules on water potential

- Solutes and ions attract water molecules to form a shell around them.
- The water molecules can no longer move around freely, so less 'free' water is present, which lowers water potential.
- The stronger a solution, the lower will be its water potential.

The effect of temperature

- As temperature increases, molecules have more kinetic energy and enzyme-controlled reactions (producing ATP) increase. However, proteins are denatured and at high temperatures both the carriers and enzymes may be inactivated.
- So active transport is affected more than diffusion by temperature.

Questions to try

Examiner's hints

- The acceptable model for a membrane is called the fluid mosaic model.
- The main components of a membrane are phospholipids. These are able to move about, giving the membrane its 'fluid' nature.
- There are relatively few proteins present
- These proteins are dotted throughout the phospholipid bilayer, giving a 'mosaic' pattern.

Q1

The generally agreed *fluid mosaic model* of cell membrane structure is not immediately clear in electron micrographs. Normally, cell membranes appear in electron micrographs to have a structure which can be shown as on the diagram below.

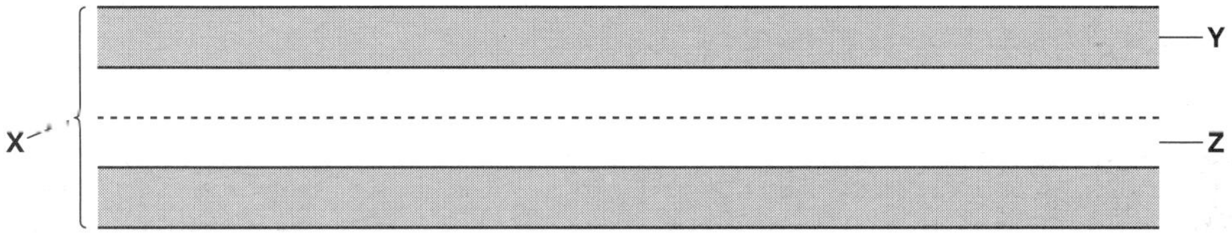

X Y Z

(a) Suggest a typical width for X (the complete cell surface membrane).

....

[1 mark]

(b) From your knowledge of the fluid mosaic model of the membrane describe:

(i) The likely composition of layer Y.

....

[2 marks]

(ii) The likely composition of layer Z.

....

[2 marks]

(c) Some substances are said to pass across cell surface membranes by *facilitated diffusion*. Explain briefly what this term means.

....

....

[2 marks]

Examiner's hints

- Particles move with a concentration gradient – diffusion – or against a concentration gradient – active transport. They never move across a concentration gradient.
- If two facts are needed for one mark you must get **both** answers correct. No half marks are awarded.
- If a question asks for four things, give only four. Do not run the risk of offering the examiner a huge list from which to choose – all your answers will be marked and any extra wrong ones could cancel out the correct ones.

Q2

(a) Give two examples of substances which pass across the membrane by simple diffusion.

....

....

[1 mark]

(b) Suggest a reason why glucose is not impeded by the hydrophobic layer of the membrane but sodium and potassium ions are.

....

....

[1 mark]

(c) Suggest how the membrane is selective with regard to the molecules that pass through it.

....

....

[2 marks]

(d) What is meant by the phrase '*move materials against the concentration gradient*'?

....

....

[1 mark]

(e) The graph shows the relationship between concentration difference across a membrane and the rate of diffusion.

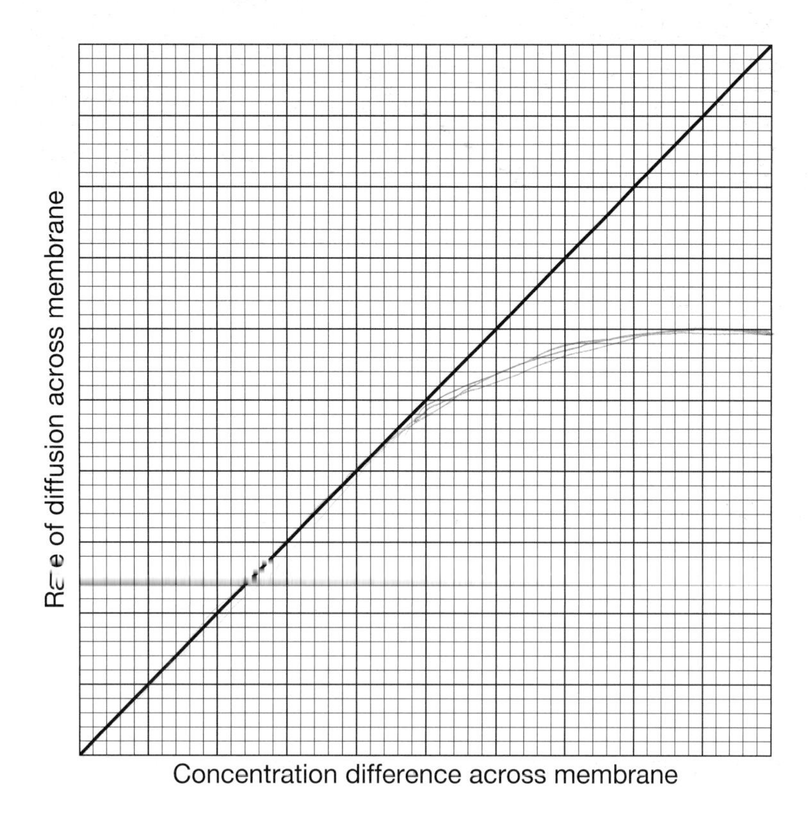

(i) On the graph sketch the corresponding curve you would expect for facilitated diffusion.

[2 marks]

(ii) Give four factors, other than the concentration gradient, which affect the rate of diffusion.

....

....

....

....

[4 marks]

Answers to Questions to try are on pp. 80 – 81.

Exam question and student's answer

The diagram shows the main stages in the cell cycle.

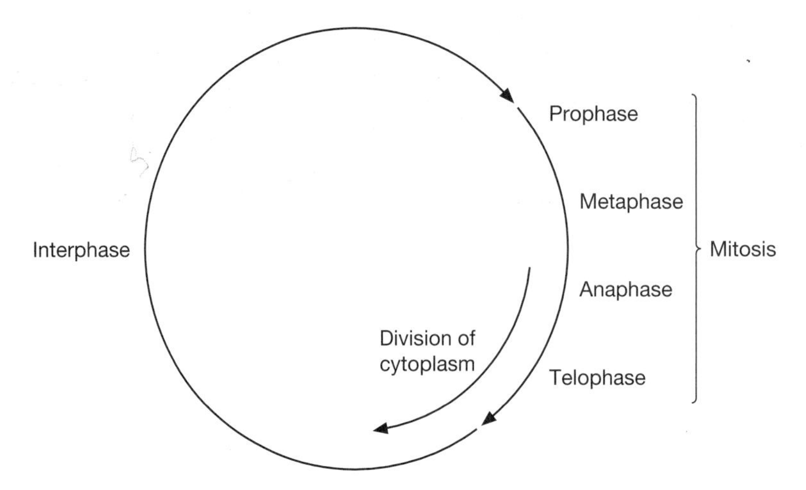

1 (a) At what stage in the cell cycle do the following events take place?

(i) Separation of daughter chromatids.

Anaphase ✓

1/1

(ii) Replication of DNA.

Prophase ✗

0/1

[2 marks]

(b) What evidence would you see on a prepared slide of a bean root tip which would show that metaphase takes longer than anaphase?

Cells in metaphase have chromosomes at the equator whereas in anaphase the chromosomes are moving apart. ✗

0/1

[1 mark]

(c) There are 44 chromatids present at the beginning of prophase of mitosis in a rabbit cell.

(i) How many *chromatids* would there be in a cell from a rabbit at the beginning of meiosis?

44 ✓

1/1

(ii) How many *chromosomes* would there be in a sperm cell from a rabbit?

22 ✗

0/1

[2 marks]

2/5

[Total 5 marks]

2 The diagrams A–E show stages of mitosis in an animal cell.

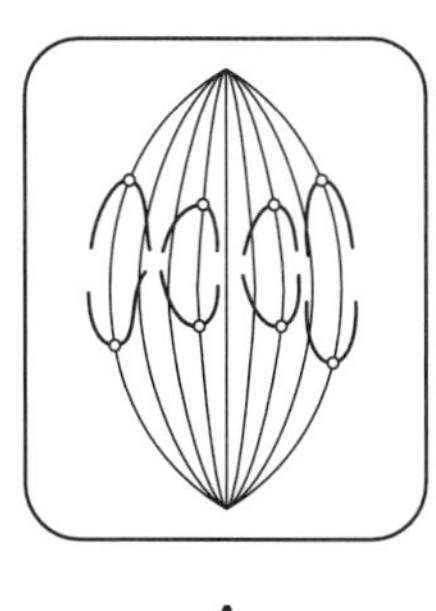
A

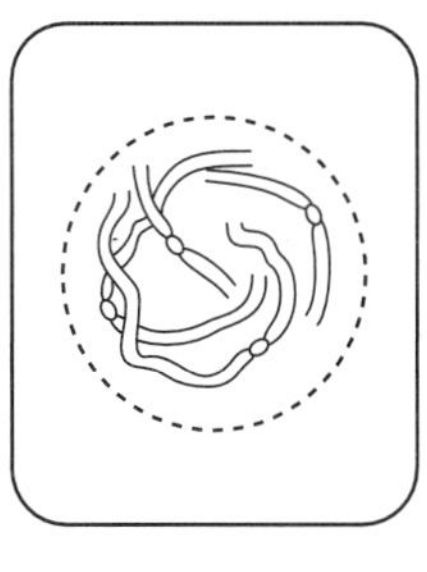
B

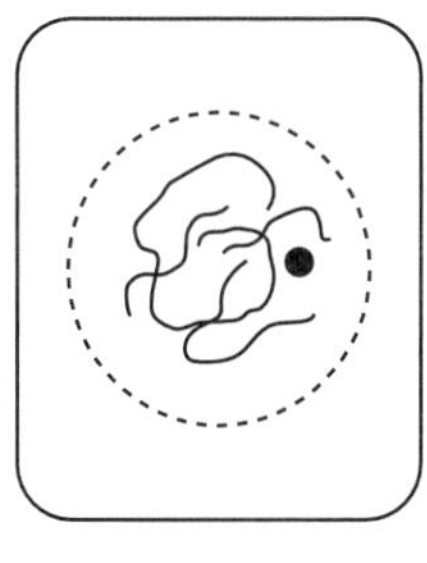
C

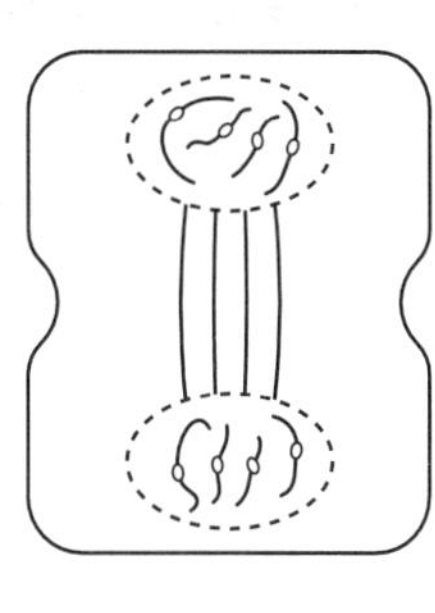
D

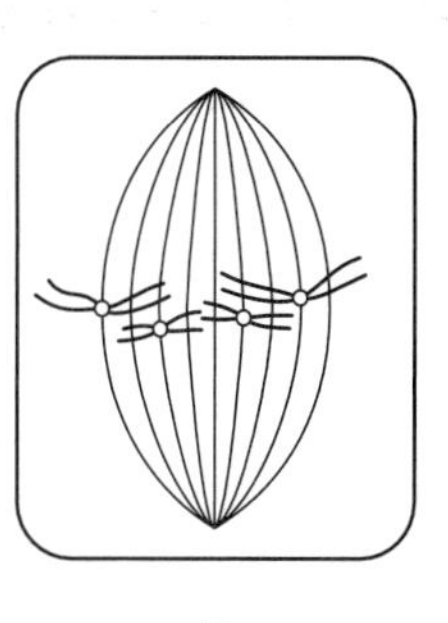
E

(a) Which of the drawings A–E shows:

(i) Anaphase

(ii) Telophase

B ✗

(iii) Metaphase

[3 marks]

(b) Name one stain that could be used to stain the chromosomes in these cells.

Iodine solution. ✗

[1 mark]

(c) The table shows the average duration of each stage in the cells of a grasshopper embryo.

Stage	Mean duration/minutes
Interphase	20
Prophase	105
Metaphase	13
Anaphase	8
Telophase	54

Give one piece of evidence from the table which indicates that these cells are dividing rapidly.

Anaphase, when the chromosomes separate, only lasts for 8 ✗ minutes.

The cell is therefore dividing very quickly.

[1 mark]

(d) Give two processes that occur during interphase and which are necessary for nuclear division to take place.

1 During interphase DNA divides. ✓

2 During interphase the centromere ✗ divides.

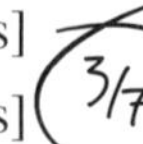

[2 marks]

[Total 7 marks] 3/7

How to score full marks

Question 1 may seem very simple but it does illustrate a number of basic principles involved with this topic and how easy it is to make mistakes.

Things you must learn

There are some things that you just have to remember. **This is an example**. Learn the cell cycle so you can remember exactly **what** happens **when**. You will either be asked to **describe what is happening** during each of the stages or **the name of the stage** – as in part (a). Replication of DNA takes place during **interphase** (the correct answer) not during the first stage of mitosis. (See 'Key points' on page 28 for some hints to make this easier to remember)

Answer the question

Part (b) asks for the **evidence** that one stage takes longer than another but the student has given a **description** of the two stages. He has wrongly read the question as 'what is the evidence that would show that a cell is in metaphase or anaphase?' The correct answer to the question is: **The number of cells visible on the slide at a particular stage indicates the time taken to complete that stage. So there would be more cells at metaphase**.

The preparation of a stained slide kills the cells. Thus the **number of cells 'frozen' during a particular stage indicates how long each stage must take**. If a stage is over very quickly, the chance of finding lots of cells at that stage is reduced.

Chromatids and chromosomes

At the beginning of mitosis and meiosis there are the same number of chromosomes as there were at the beginning of interphase, but twice the amount of DNA because each chromosome is made up of two chromatids. in part (c) (i) the student has correctly given the number of chromatids.

But:

- **You must understand the difference between a chromosome and a chromatid** and always be sure you **use the correct term**.
- When DNA replicates, two chromatids are formed but **they remain attached together at the centromere as one chromosome**. So a chromosome can be made up of a single strand of DNA (the situation at the end of mitosis or meiosis) or of two strands joined together (the situation at the beginning of mitosis or meiosis).
- The student has confused chromosome with chromatid in part (c) (ii). The process of meiosis **halves the number of chromosomes** found in a 'body' cell, but if the body cell has 44 chromatids it must have 22 chromosomes. So the answer should be half of 22 (i.e. 11), not half of 44.

How to score full marks

Look for patterns

As with the first question, in question 2 **you need to have learnt how to recognise the stages of mitosis by the positions of the chromosomes.** In part (a) the student has correctly recognised the diagrams that show anaphase and metaphase. However, during telophase the chromosomes, at either pole of the cell, group together and a new nuclear membrane is formed. **The diagram that shows telophase is therefore D.**

Link practical and theory

- You must check your syllabus to see where practical techniques could be asked for in theory papers. Food tests and staining are examples.
- Iodine solution, which the student gave in part (b), is a stain, and will show up the nucleus. However, **to see detail of chromosomes a special stain – acetic orcein – is needed.**

Read the data

Try to relate any data given in a table to your theory. In part (c), the student has simply spotted the shortest time period in the table and restated the question. A correct answer would be along the lines of '**The shorter interphase is, the sooner the nucleus divides again and thus the faster the cells are dividing. The data show that the cells divide every 20 minutes.'**

Beware of similar words

To prepare the cell for mitosis the **centriole** divides. These structures migrate to the two poles of the cell and form the spindle. In part (d), the student has confused the **centromere** – the structure holding the chromatids together and responsible for their attachment to the spindle – with the **centriole**.

Don't make these mistakes ...

Each type of nuclear division is designed for a different function and you should know the difference:

- Mitosis produces identical cells for growth whilst meiosis produces gametes for sexual reproduction.
- 'My toes (mitosis) grow' may help you remember their roles.

Mitosis and meiosis are words that look alike. Be sure you know how to spell them. If they are not spelt exactly right you could lose marks.

Interphase is **not** one of the stages of mitosis.

Key points to remember

Stages of mitosis

Stage of mitosis	Mnemonic to remember the stages	Word to remember what is happening	How to identify the stage	Events taking place
Prophase	**P**lastic	**P**resent	Chromosomes visible, each consisting of a double structure made up of identical chromatids joined at centromere	Nuclear envelope starts to break down. Chromosomes move to equator.
Metaphase	**M**eat	**M**iddle	Chromosomes arranged across the middle of the cell (equator)	Chromosomes attached to protein fibres that form the spindles
Anaphase	**A**in't	**A**part	Chromatids are separated	Chromatids move apart, pulled by the contraction of spindle fibres, to opposite poles of the cell. *Once chromatids have separated from each other they are called chromosomes
Telophase	**T**asty	**T**wo	Two groups of chromosomes at opposite ends of the cell	Chromosomes start to uncoil, lose their distinct appearance and a nuclear envelope forms

Seeing the chromosomes

- A special stain must be used – acetic orcein.
- This colours the chromosomes purple and enables their pattern to be seen.

What you do	Why
Remove the **tip** of a root (onion or garlic)	This is the growing region where mitosis is occurring
Place on a watch glass with **hydrochloric acid** and **acetic orcein**	The acid separates the cells while the acetic orcein stains the chromosomes
Remove the last 2 mm of the root, place it on a slide and squash it	This produces a single layer of cells so that the pattern of the chromosomes is visible

Interphase

- This is the stage when the nucleus is not dividing.
- It is not a stage of mitosis or meiosis.
- During interphase the amount of DNA in a cell doubles – the DNA is said to **replicate**.

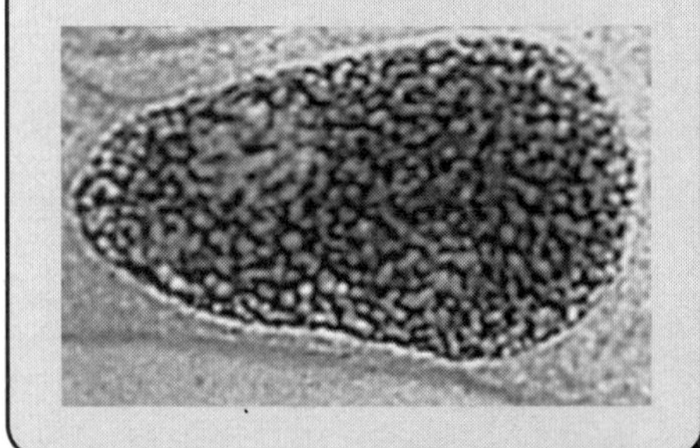

Meiosis

- Meiosis is also the division of the nucleus.
- It occurs in two stages: meiosis I and meiosis II.
- Following each nuclear division the cytoplasm divides.
- Four cells are formed that each have half the number of chromosomes of the parent cell.

MEIOSIS I
First cytoplasmic division
Homologous chromosomes separate

MEIOSIS II
Second cytoplasmic division
Chromatids separate

Result

Half the number of chromosomes

Purpose

Produce gametes

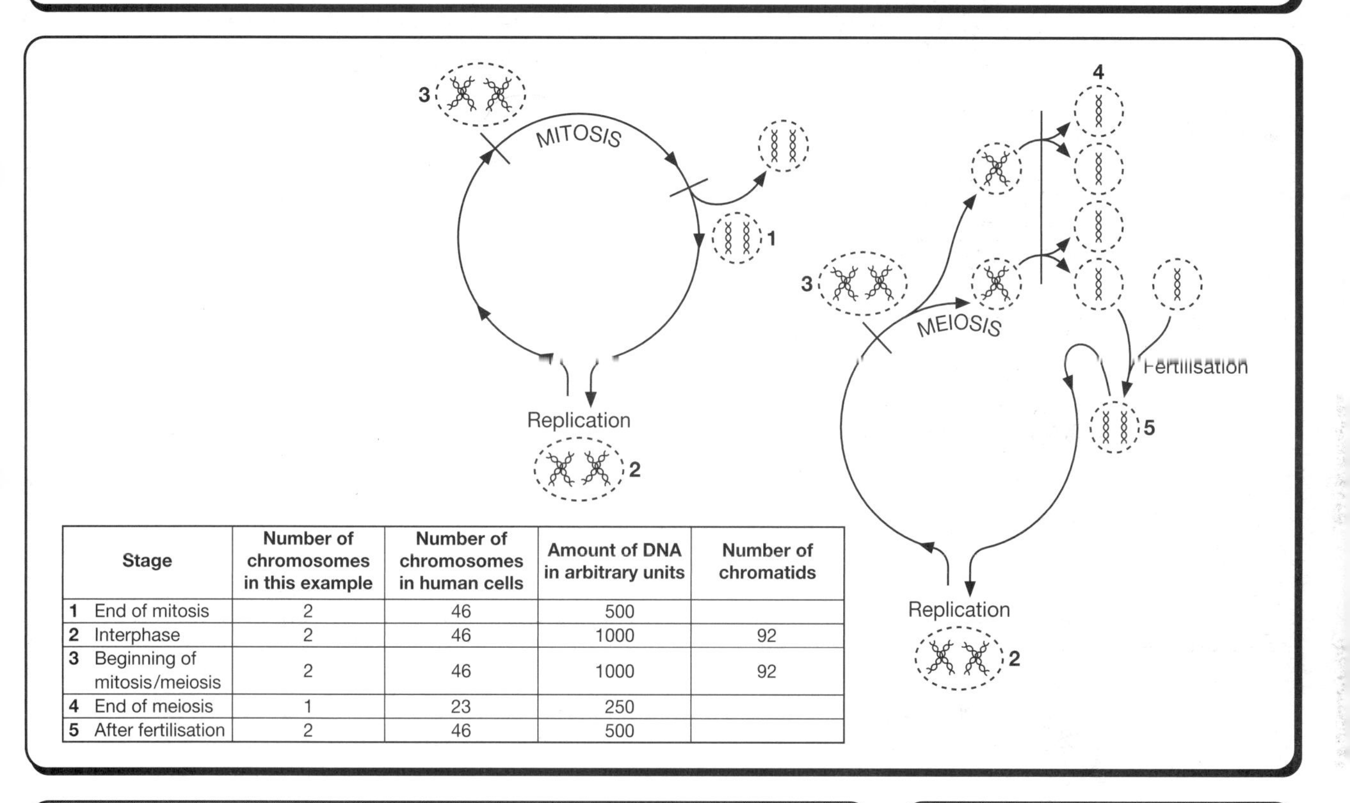

Stage	Number of chromosomes in this example	Number of chromosomes in human cells	Amount of DNA in arbitrary units	Number of chromatids
1 End of mitosis	2	46	500	
2 Interphase	2	46	1000	92
3 Beginning of mitosis/meiosis	2	46	1000	92
4 End of meiosis	1	23	250	
5 After fertilisation	2	46	500	

Cell cycle

There are five main stages.

- **Interphase**

1 Growth phase: $\mathbf{G_1}$. Protein is synthesised and the cell grows. Cell organelles increase in number.

2 DNA synthesis phase: **S**. DNA replicates.

3 Second growth phase: $\mathbf{G_2}$. Occurs immediately before mitosis or meiosis. Proteins required for cell division are synthesised. Centrioles divide.

- **Mitosis**

4 Nucleus divides.

- **Cytokinesis**

5 Cytoplasm divides.

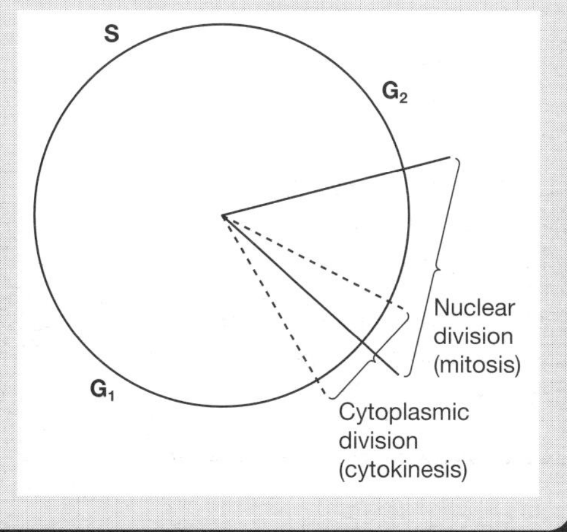

Mitosis

- Mitosis is the division of the nucleus that takes place during the growth and development of an organism.
- The cytoplasm divides once.
- Two cells are formed that have identical genetic information to the parent cell, containing the same number of chromosomes and therefore the same amount of DNA.
- It is divided into a number of stages

Questions to try

Examiner's hints

- If a diagram shows paired structures separating – i.e. overlapping 'V' shapes on spindles within cells (see figure, right) – then the diagram shows meiosis.
- Centrioles are the organelles that form the spindle fibres and are found at the ends or poles of the dividing cell.
- 'Centri**oles** at the **poles**' may help you distinguish them from centromeres, the structures that hold chromatids together and attach the chromosome to the spindle fibres.

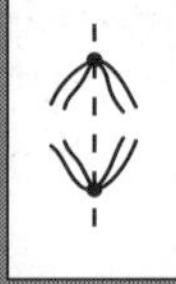

Q1

(a) When a cell divides, the genetic material can divide by mitosis, by meiosis or by neither of these processes. Complete the table with a tick to show the process by which you would expect the genetic material to divide in each of these examples.

	Mitosis	Meiosis	Neither
The division of plasmids in bacterial reproduction			
The stage in the formation of male gametes in a plant in which haploid daughter cells are formed from a haploid parent cell			
Cell division which takes place in the growth of a human testis between birth and five years of age			
The stage in the life cycle of a protoctistan in which a large number of genetically different spores are produced			

[2 marks]

(b) The diagram shows a cell during the first division of meiosis.

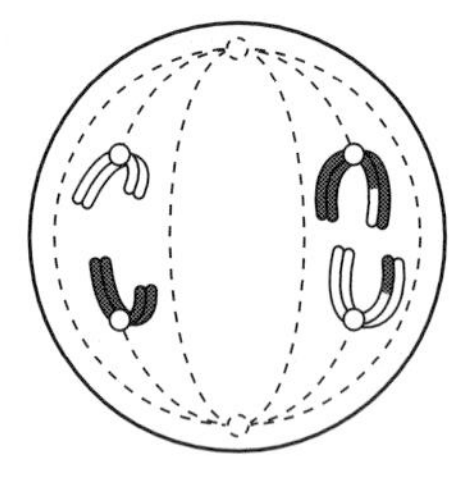

Complete the diagram to show the appearance of the chromosomes in each of the four daughter cells formed at the end of the second division of meiosis.

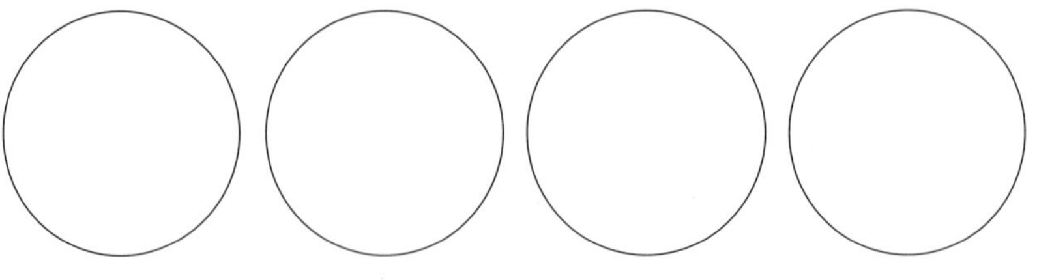

[2 marks]

(c) In an insect, 16 chromatids were visible in a cell at the start of the first division of meiosis. How many chromosomes would there be in a normal body cell from this insect?

....

[1 mark]

Examiner's hints

- Don't be put off if the diagram in a question has additional material or is not familiar. What is being tested is always your knowledge of the principles found in your syllabus.
- Whenever 'maths' is involved in AS question it will be straightforward. Think through the maths carefully.

The diagram summarises the events of the cell cycle.

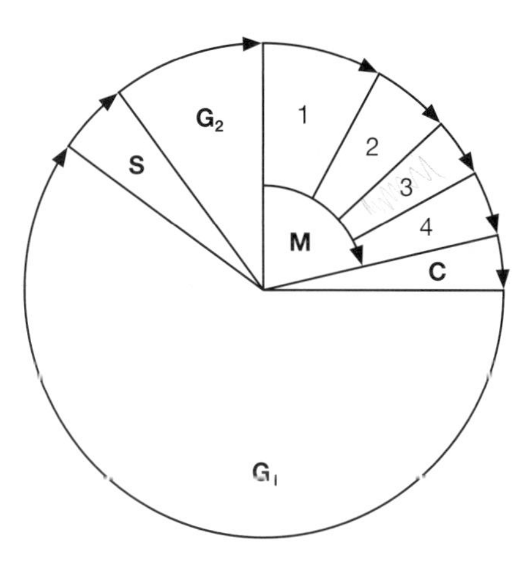

Key
M Mitosis:
1 prophase
2 metaphase
3 anaphase
4 telophase
C cytoplasm divides

Interphase:
G_1 } growth phases
G_2
S replication of DNA

(a) Shade the diagram to show where the spindle fibres would be shortening.

[1 mark]

(b) (i) During the G_1 phase, the ratio of nucleus diameter to cell diameter decreases. Explain why.

....

....

[1 mark]

(ii) How do the events of the G_1 phase prepare the cell for mitosis?

....

....

[1 mark]

(c) Describe how the arrangement of chromosomes during metaphase of mitosis differs from the arrangement of chromosomes during anaphase.

....

....

[2 marks]

Examiner's hints

- Mitosis takes place in every organ of an animal's body. However, it occurs at the tip of the root and shoot in young flowering plants.
- Meiosis occurs only in the sex organs:
  - ovaries and testes of animals
  - ovaries and anthers of flowering plants.
- Meiosis does not occur in the gametes (e.g. sperm cells and egg cells). These are the products of meiosis.
- To recognise the stage shown in diagrams or to put a series of diagrams in order, remember the mnemonic **P**lastic **M**eat **A**in't **T**asty.

Q3

(a) Explain why root tips are particularly suitable material to use for preparing slides to show mitosis.

....

....

[1 mark]

(b) Give a reason for carrying out each of the following steps in preparing a slide showing mitosis in cells from a root tip.

(i) The tissue should be stained.

....

....

[1 mark]

(ii) The stained material should be pulled apart with a needle and gentle pressure applied to the cover slip during mounting.

....

....

[1 mark]

(c) The drawing has been made from a photograph showing a cell undergoing mitosis.

In which stage of mitosis is the cell shown in this drawing?

....

[1 mark]

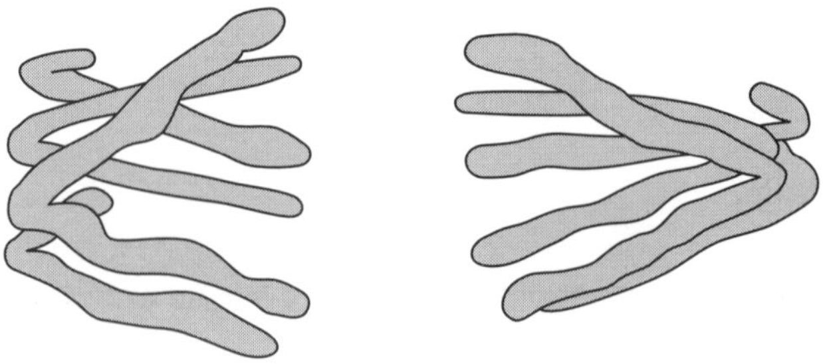

Answers to Questions to try are on pp. 82 – 83.

Exam question and student's answer

The diagram below represents two nitrogenous bases from a sample of a nucleic acid.

(a) The right-hand molecule is thymine. Identify the left-hand molecule.

Thymine always pairs with the same base so the left hand molecule is Adenine. ✓

[1 mark]

(b) Suggest which nucleic acid is most likely to have been used to supply the sample.

Adenine and Thymine are bases found in DNA ✓ so that would be the source of these bases.

[1 mark]

(c) Name and place into its chemical group one other nitrogenous base that you would expect to find in this sample of nucleic acid.

Guanine ✓ is another nitrogenous base found in DNA, it is a pyrimidine.

[2 marks]

(d) State:

(i) what is represented by the dotted lines on the diagram

Bonds. ^

(ii) their importance in the structure of this nucleic acid

They hold the nitrogenous bases together. ✗

[2 marks]

(e) (i) Name two components of this nucleic acid, other than nitrogenous bases.

Phosphate ✓ and ribose sugar. 1/2

(ii) Name the type of reaction that bonds the bases to these other compounds to make the complete macromolecule.

The reaction is called a Hydrolysis. 0/3

[3 marks]

[Total 10 marks] 4/10

How to score full marks

Recall

A great deal of the AS exam will involve recalling facts – look for terms like 'state,' 'identify' and 'name', as in this example.

Purines and pyrimidines

If the name of the nitrogenous base has a Y in, then it is a pyrimidine. In part (c) the student correctly gave guanine as another base found in DNA – but as there is no Y in the name it must be a purine, not a pyrimidine.

'A'-level quality

There are many types of bond that hold molecules together. In part (d) (i) the bonds joining the nitrogenous bases are **hydrogen bonds**. You would be expected to know that and **the answer 'bond' would not be enough**.

The student's answer to part (d) (ii) is correct but incomplete. The **student has not given the importance of these bonds compared with any other in the DNA molecule**. The correct answer is that **hydrogen bonds are weak and are thus easily broken**, allowing the DNA strands to separate during replication or transcription.

Don't lose track

The student has done quite well in parts (a) and (b) but in part (e) has forgotten that the rest of the question is referring to DNA and gives the pentose sugar as ribose instead of **deoxyribose**. He also muddles hydrolysis (breakdown of a molecule by the addition of a molecule of water) with condensation (the joining of two molecules by the removal of a molecule of water) – the correct answer.

Don't make these mistakes ...

Many biological molecules are joined by condensation reactions and broken down by hydrolysis. You will have learned that proteins, lipids and carbohydrates are joined and separated by these reactions and may be expected to draw the conclusion that nucleic acids are affected in the same way. Although there does not have to be any synopsis – linking of areas of the syllabus – in the AS exam, you can't avoid relationships in biology.

If you are asked to 'name' or 'state', there is no need to put the name into a sentence. It will not lose you marks, but doing so won't give you any more marks and will waste time.

Key points to remember

Nucleic acids

Structural component	DNA	mRNA	tRNA
Deoxyribose sugar	✓		
Ribose sugar		✓	✓
Guanine, cytosine, adenine	✓	✓	✓
Thymine	✓		
Uracil		✓	✓
Phosphate	✓	✓	✓

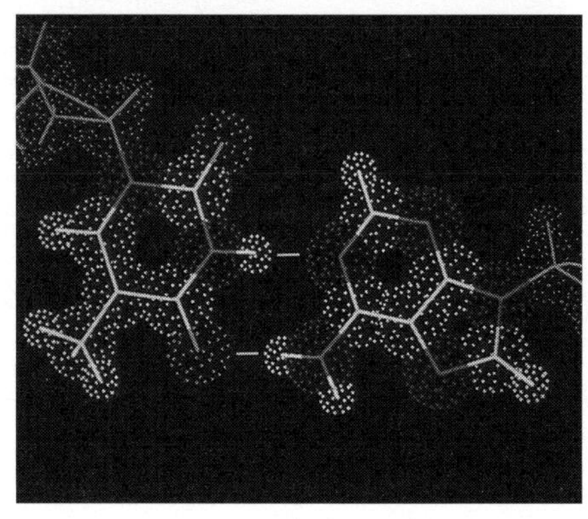

Computer-generated image of base-pairing in DNA.

Structure of DNA and RNA

Step 1

Nucleotides of DNA and RNA

DNA

RNA

Consist of:

Phosphate

Purine

Pyrimidine

Nitrogenous bases

Deoxyribose sugar

Ribose sugar

Step 2

Polynucleotides – nucleotides linked by condensation reaction

Step 3

DNA – Two chains link through complementary bases, joined by hydrogen bonds

RNA – Remains either as single strand or folds, some sections linked by complementary bases

mRNA

Codon

Two polynucleotide chains – anti-parallel

Straight chain

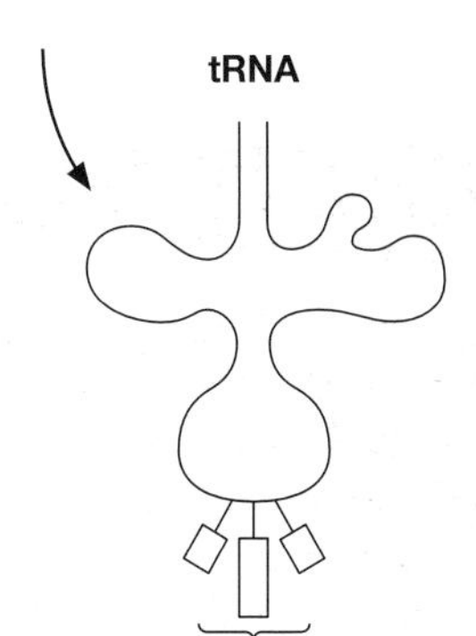

Folded chain

Key points to remember

Function of DNA

- Codes for a polypeptide or protein.
- Consists of two polynucleotide strands but only one, the **sense strand**, carries the code.
- The sense strand is the sequence of bases that forms the genetic code.
- The other, non-coding, strand is the **anti-sense strand**. It consists of a complementary base sequence which is important only during replication.

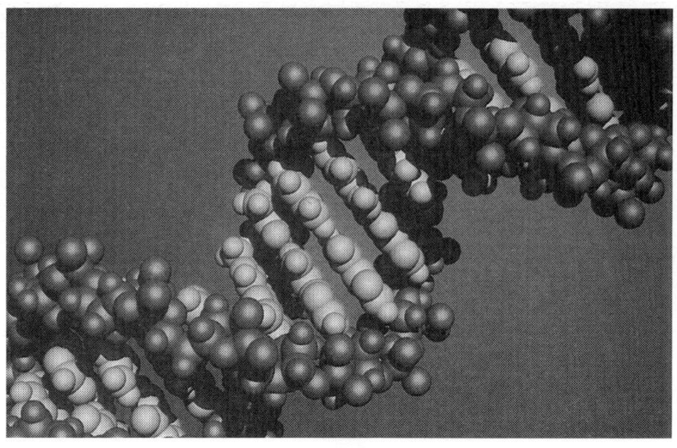

Computer-generated model of DNA

Function of mRNA

- It is a **complementary copy** of the genetic code.
- mRNA acts as a **messenger** molecule.
- It is small enough to leave the nucleus and carry the code to a ribosome in the cytoplasm.
- Here protein synthesis takes place.

Function of tRNA

- A specific tRNA molecule links with a specific amino acid in the cytoplasm.
- If the **anticodon** of the tRNA is complementary to the **codon** of the mRNA in the ribosome the amino acid it holds will be added to the existing chain.

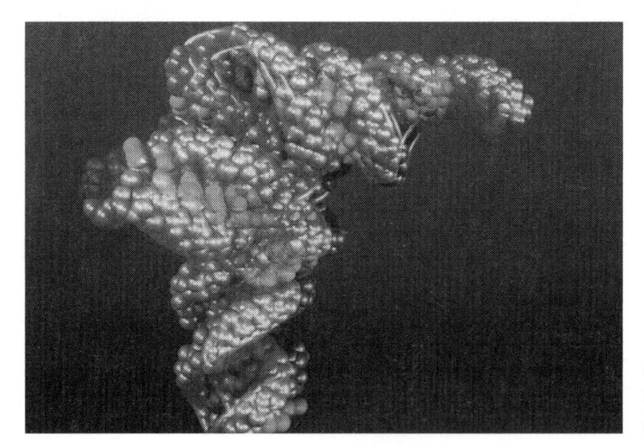

Computer-generated model of the three-dimensional structure of tRNA

Genetic code	
Triplet	Although there are four bases they are read in groups of three – e.g. ATT, CCG – thus producing 64 possible codes
Degenerate	There are more combinations of bases to produce codes (64) than amino acids (20), so several triplets code for the same amino acid – e.g. CCA and CCC both code for the amino acid glycine. The first two bases are often more important than the third
Non-overlapping	An individual base does not occur in more than one triplet. For example, if the sequence of bases was ACCGGT … then it would be read ACC/GGT not ACC/CGG
Universal	The same triplet of bases always codes for the same amino acids in every organism

Questions to try

Examiner's hints

- Don't expect the same diagram always to be used to represent DNA. Use the molecules that link the strands together as a key – these are the nitrogenous bases. They are attached to the sugar, which in turn is joined to the phosphate molecule.
- Hydrogen bonds between the bases are weak and allow the strands of DNA to separate easily for both replication and transcription.
- In many questions the letters A, T, C, and G are used to represent the names of the nitrogenous bases. You must remember their names in case you are asked.

The diagram below represents a two-dimensional view of part of a **DNA** molecule. Study the diagram and then answer the following questions.

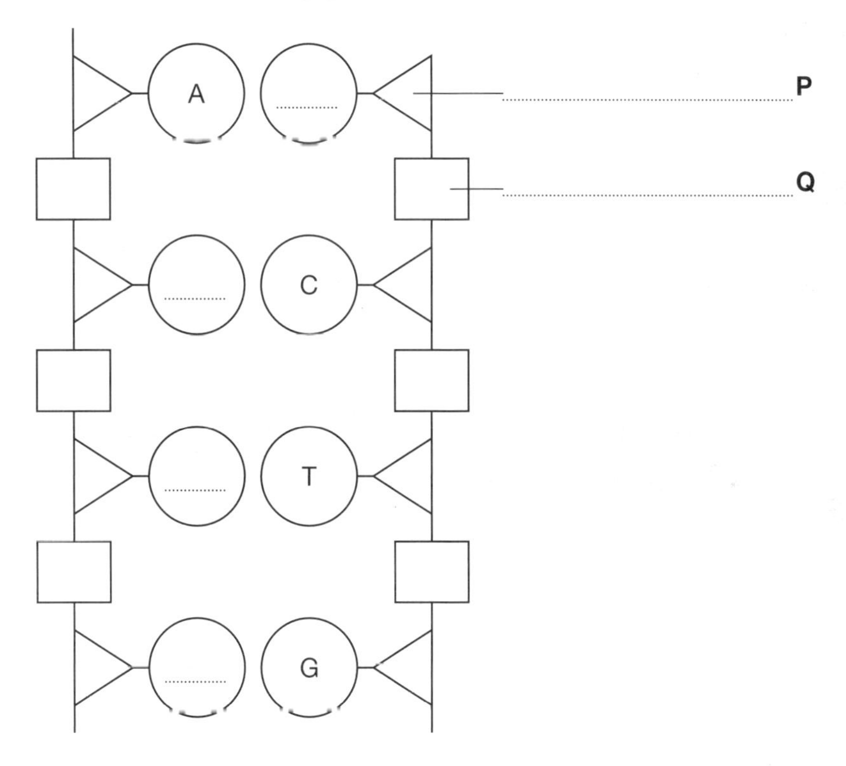

(a) Label the parts of the diagram at P and Q.

[2 marks]

(b) The circles represent organic bases, some of which are labelled. Mark **on the diagram** the appropriate identity of the unlabelled bases, using the appropriate letters.

[1 mark]

(c) Which of the bases is **not** present in RNA?

....

[1 mark]

(d) How are the bases in the complementary strands of **DNA** held together?

....

[1 mark]

(e) Briefly describe the process of DNA replication.

....

....

....

....

[3 marks]

(f) In a sample of **DNA** it is possible to measure the relative proportions of the following bases. Based on what you know of the structure of DNA, which **two** of the following formulae are most likely to be correct?

(i) A + G = T + C **(ii)** A + T = G + C **(iii)** A + C = G + T

[1 mark]

Examiner's hints

- There are two types of nitrogenous base in nucleic acids – purines and pyrimidines. Although they are both nitrogenous bases, purines are twice as big as pyrimidines. The smallest word is the longest molecule.
- An easy way to remember which of the named bases is which type is given in the table below.

P**y**rimidine	Th**y**mine C**y**tosine
Purine	Adenine Guanine

- If the name of the base contains the letter **Y** then it is a p**Y**rimidine.

The following diagram represents part of a DNA molecule and a mRNA molecule.

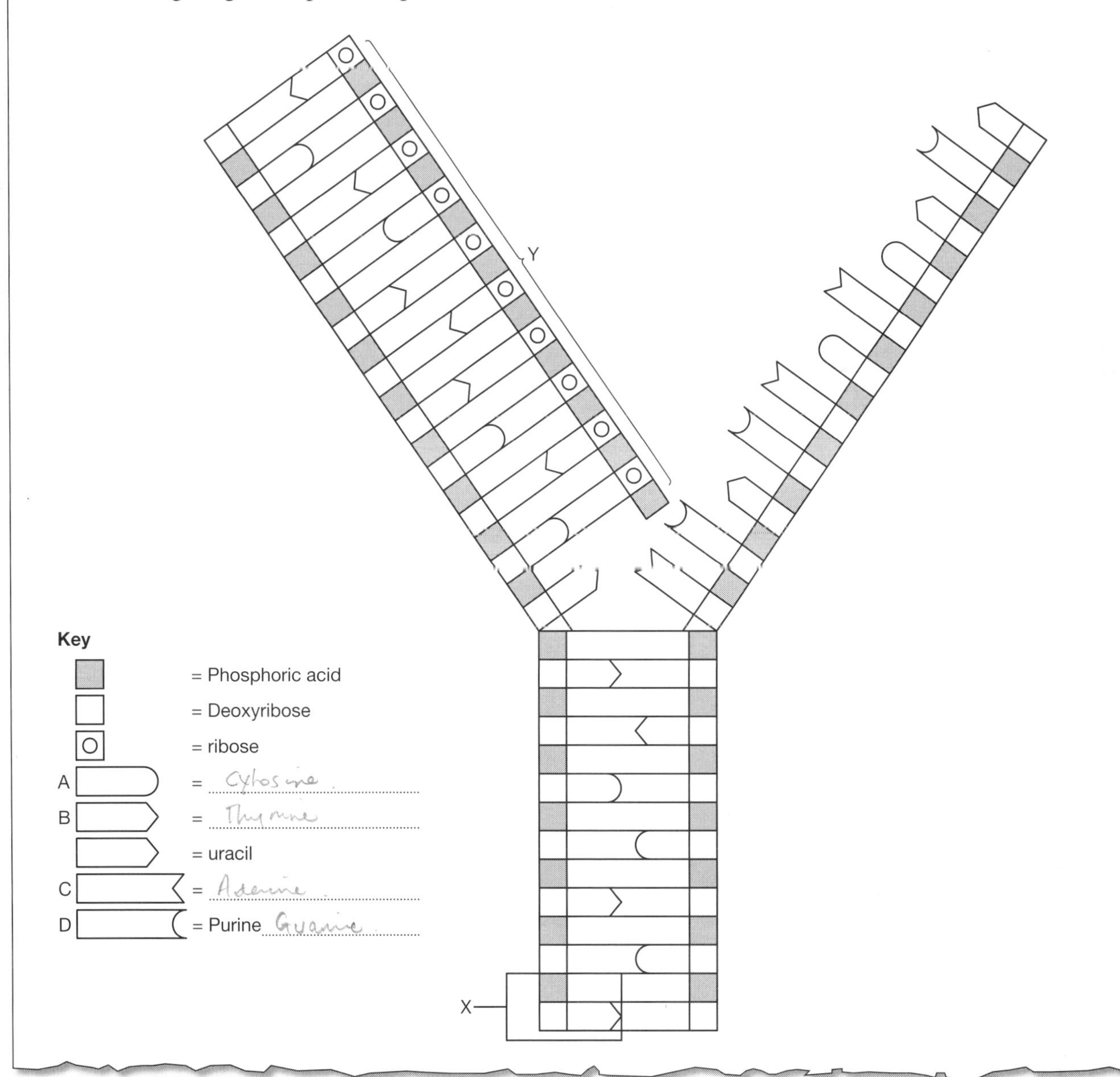

(a) Name molecules **A**, **B, C** and **D** in the spaces provided in the key.

[4 marks]

(b) (i) What is the name given to the molecular sub-unit shown in the box **X**?

....

[1 mark]

(ii) Give the names of the component molecules which make up this structure.

....

....

....

[3 marks]

(c) On the diagram, use a **ruled guide line** and the letter H to show the position of a hydrogen bond.

[1 mark]

(d) Condensation reactions are involved in the production of DNA. Give the names of **two pairs** of molecules which are linked in such a way.

....

....

[2 marks]

(e) Give **two** pieces of evidence from the diagram which indicate that the molecule Y is RNA and not half a strand of DNA.

(i)

....

....

(ii)

....

....

[2 marks]

Answers to Questions to try are on pp. 84 – 86.

Exam question and student's answer

The table shows the mRNA base sequences (codons) which code for specific amino acids. The names of the amino acids have been abbreviated.

UUU	Phe	UCU	Ser	UAU	Tyr	UGU	Cys
UUC	Phe	UCC	Ser	UAC	Tyr	UGC	Cys
UUA	Leu	UCA	Ser	UAA	stop codon	UGA	stop codon
UUG	Leu	UCG	Ser	UAG	stop codon	UGG	Trp
CUU	Leu	CCU	Pro	CAU	His	CGU	Arg
CUC	Leu	CCC	Pro	CAC	His	CGC	Arg
CUA	Leu	CCA	Pro	CAA	Gln	CGA	Arg
CUG	Leu	CCG	Pro	CAG	Gln	CGG	Arg
AUU	Ile	ACU	Thr	AAU	Asn	AGU	Ser
AUC	Ile	ACC	Thr	AAC	Asn	AGC	Ser
AUA	Ile	ACA	Thr	AAA	Lys	AGA	Arg
AUG	Met	ACG	Thr	AAG	Lys	AGG	Arg
GUU	Val	GCU	Ala	GAU	Asp	GGU	Gly
GUC	Val	GCC	Ala	GAC	Asp	GGC	Gly
GUA	Val	GCA	Ala	GAA	Glu	GGA	Gly
GUG	Val	GCG	Ala	GAG	Glu	GGG	Gly

(a) (i) Give the amino acid coded for by the DNA base sequence GGT.

CCU Pro ✓ (1/1)

(ii) Give one possible tRNA base sequence for the amino acid Tyr.

ATA (0/1)

[2 marks]

(b) Describe how cells use the base sequence of a molecule of DNA to produce a polypeptide.

The strand of DNA that is needed unzips, ✓ free RNA nucleotides ✓ come in and line up against the sense strand for DNA according to the pairing rule A–U, ✗ G–U, C–G, T–A. The RNA is then joined together to form mRNA and leaves the nucleus. It enters the cytoplasm where it is met by ribosomes. tRNA picks up specific amino acids, tRNA then lines up against the mRNA doing this it brings in amino acids which are on the anticodons, they match the mRNA codons. The amino acids are joined together by peptide bonds and begin to form a polypeptide.

4/7

[7 marks]

(c) Sometimes errors occur during the copying of a sequence of bases. Use the information given to explain why some errors have less severe consequences than others.

If a substitution occurs then this is where one of the bases in the sequence is changed for another one. For example if the U in GU(U) was substituted for an A then it will still code for the amino acid Val.

[3 marks]

[Total 12 marks]

How to score full marks

Triplets – codons – anticodons

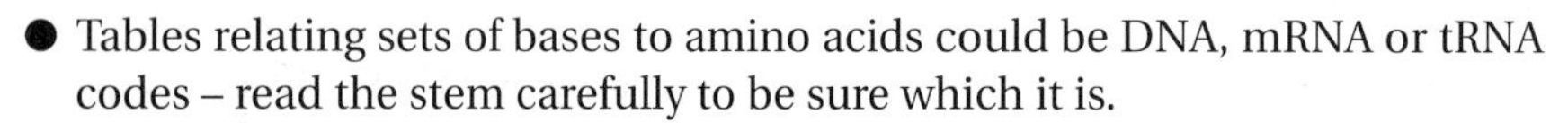

- Tables relating sets of bases to amino acids could be DNA, mRNA or tRNA codes – read the stem carefully to be sure which it is.

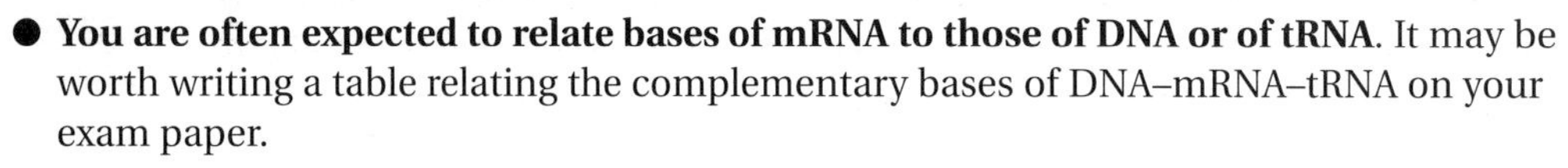

- **You are often expected to relate bases of mRNA to those of DNA or of tRNA**. It may be worth writing a table relating the complementary bases of DNA–mRNA–tRNA on your exam paper.

DNA	mRNA	tRNA
A	U	A
T	A	U
C	G	C
G	C	G

Basic confusion

- **Remember that the base thymine in DNA does not appear in RNA**. It is replaced as a complement to adenine by uracil. So AUA is the correct answer to part (a) (ii).
- **If you are going to give examples of base pairing then they must all be correct** – G and U are not complementary. Note therefore that the mark given in part (b) for the correct example of base pairing (A–U) is cancelled when an incorrect pairing (G–U) is given.

Structure of tRNA

- **The anticodon is situated at one end of the tRNA molecule while the amino acid attachment site is at the other end**. The student didn't make that clear. She wrote that the amino acid is on the anticodon.

How to score full marks

- **Try to write short sentences** so there is no doubt (as there is in part (b)) that codons/anticodons and not amino acids/codons are linked.

Tell the whole story

- When part of a question is worth a lot of marks, like this is (7), then you know you must **tell the whole story**. It's like giving directions to someone – if you miss out a landmark or a change of direction the person will get lost. It's the same in these longer questions
- **Marks may be awarded for an idea or two in a sequence. If you miss anything out you could lose the mark**. Questions often have 10 possible marks, up to a maximum of 5. The example below shows each step in part (b) - it tells the whole story!!

- DNA is transcribed to form messenger RNA.
- The DNA molecule unzips (by breaking the hydrogen bonds between the nitrogenous bases) and one strand, the sense strand, is used as a template.
- RNA nucleotides complementary to the bases of the sense strand are assembled in sequence and are joined together to form mRNA.
- This leaves the nucleus via the nuclear pores and in the cytoplasm associates with a ribosome.
- Specific tRNA molecules each combine with a specific amino acid.
- A triplet of bases at the other end of the tRNA, called an anticodon, matches complementary triplets of bases on the mRNA – called the codon.
- This determines which amino acid is placed where in the sequence.
- The process is repeated using the next codon of mRNA and the amino acids join by peptide bonds.
- A whole chain of amino acids forms a polypeptide.

Give the theory behind the example

In part (c), students were expected to mention the fact that this is a **degenerate code and that changing the last of the triplet of bases may have no effect**. They were then expected to **give an example** from the table. For example, UCU might be the codon, which codes for serine. Changing U to C, A or G would still code for serine.

Don't make these mistakes ...

DNA and RNA are different molecules made of different nucleotides – **you cannot make RNA from the nucleotides for DNA**.

DNA is the code for the production of protein – **DNA is not made up of protein**.

Don't confuse the processes of transcription and translation. Transcription (copying) takes place in the nucleus (bases on DNA are copied as bases on mRNA); **translation** (changing) takes place within the ribosomes in the cytoplasm (bases on mRNA are used to form the sequence of amino acids making up a polypeptide).

Key points to remember

Protein synthesis

There is a direct relationship between the base sequence of DNA and the sequence of amino acids that makes up a protein.

DNA →	mRNA →	tRNA →	protein
triplet	codon	anticodon	amino acid
CCA	GGU	CCA	glycine
AAA	UUU	AAA	phenylalanine
GTG	CAC	UCG	histidine

- Protein synthesis begins with a code – DNA (**triplets** of bases in a line). Only the **sense** strand is used.

Transcription

- Transcription of DNA now occurs. **Transcribe means copy** – using DNA as a template, complementary RNA nucleotides are joined to make mRNA.
- mRNA can leave the nucleus. Each triplet code of bases from DNA is now represented by three bases on mRNA, called a **codon**.

Translation

- Translation of mRNA now occurs. **Translate means to interpret** – using mRNA as a template, complementary tRNA molecules (which are joined to specific amino acids) bring amino acids to the mRNA strand in a sequence determined by the codons.
- Parts of the tRNA complementary to a codon and known as the **anticodon** allow the correct tRNA molecule to position its amino acid in the right order to create the polypeptide.

Summary of transcription and translation

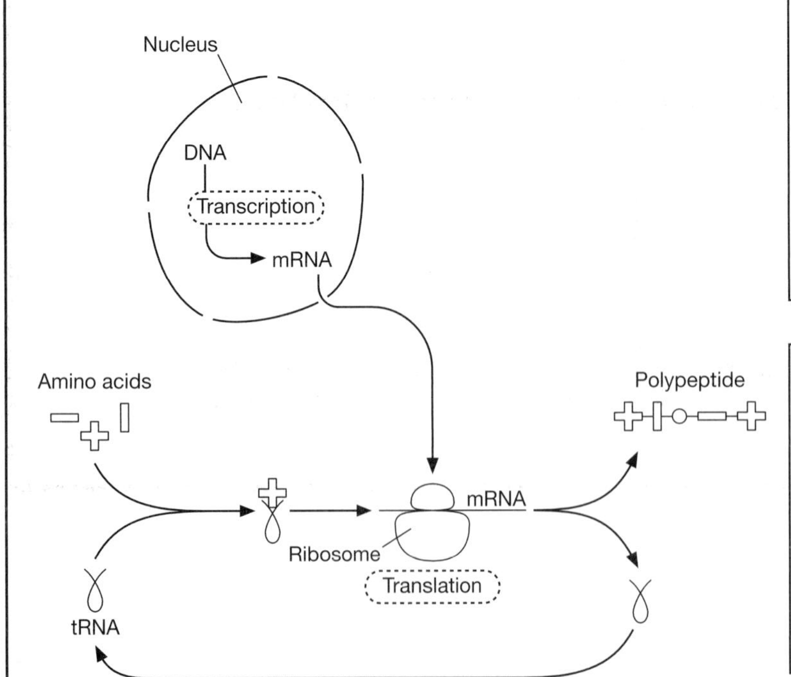

Transcription

- Hydrogen bonds between DNA strands are broken
- Sense, or coding, strand is exposed
- RNA nucleotides line up against complementary bases
- RNA polymerase joins nucleotides
- mRNA formed
- mRNA leaves the nucleus, moving into the cytoplasm

Translation

- mRNA associates with the ribosome
- tRNA brings specific amino acids to the codon
- Anticodon of tRNA matched to complementary codon of mRNA
- Peptide bonds formed between amino acids
- Polypeptide forms
- tRNA returns to the cytoplasm

Genetic engineering

1 First **isolate the gene.** There are three main ways of doing this:

1	2	3
The base sequence that forms the gene can be located using a **DNA probe**. This is a single strand of DNA, which can bind to the complementary base sequence and identifies the required gene. **Restriction enzymes** are used to cut the DNA at a specific base sequence on either side of the gene, producing **sticky ends**.	The gene can be produced from the relevant mRNA molecules. An enzyme known as **reverse transcriptase** is used to make a single strand of complementary DNA, or **cDNA**. From this a double-stranded DNA molecule of the gene can be produced.	It is possible to work backwards from the amino acid sequence that you want to produce to synthesise a piece of DNA with the appropriate base sequence.

2 Once you have isolated the gene from the protein you wish to produce, the next stage is to **introduce it into a suitable host**, such as a bacterium.

For this a **vector** is used, most commonly a **plasmid**, a circular piece of DNA found inside a bacterium. The plasmid is cut open at the same base sequence using the *same* restriction enzyme that was used to cut the gene from the original DNA. This results in cut ends (called **sticky ends**) with the same sequence of bases and allows the gene to join up with the DNA in the plasmid. Another enzyme, **DNA ligase**, is used to join these pieces of DNA together. The plasmid is now known as a **recombinant** plasmid.

3 The next step is to **get the recombinant plasmid into the bacterial cells.**

4 The bacterial cells containing the required gene are now **grown in a fermenter**. All of the cells containing the required gene will produce the required protein product.

Key points to remember

Genetic fingerprinting

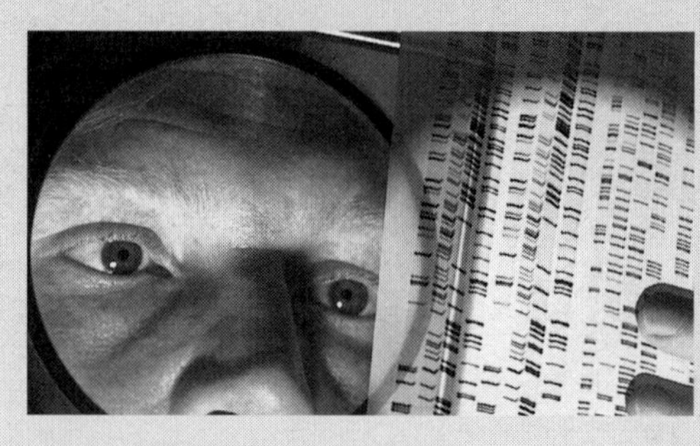

- This is a technique used to compare samples of DNA.
- DNA contains non-coding sequences; these sequences of bases may be repeated.
- The number of repeats varies from person to person, and is unique to each individual.
- This technique consists of the following steps:

1. DNA is isolated from a suitable sample, such as blood or semen.
2. The DNA is cut into smaller pieces using **restriction enzymes**. Some of the pieces will contain the repeated sequences that are being investigated – if there are few repeats the piece of DNA concerned will be small, many repeats and the DNA segment will be longer.
3. **Electrophoresis** is used to separate the DNA pieces. They separate due to both size and electrical charge. The smaller pieces, with few repeats, will travel further than longer ones, with many repeats, forming different bands of DNA.
4. The bands of DNA containing the repeated sequences are now identified using a **DNA probe** for those sequences.
5. The position of the bands can be used to compare samples from different individuals.

Polymerase chain reaction (PCR)

- This enables large amounts of DNA to be produced from very small samples by copying it many times.
- It is used to produce large amounts of DNA for analysis.

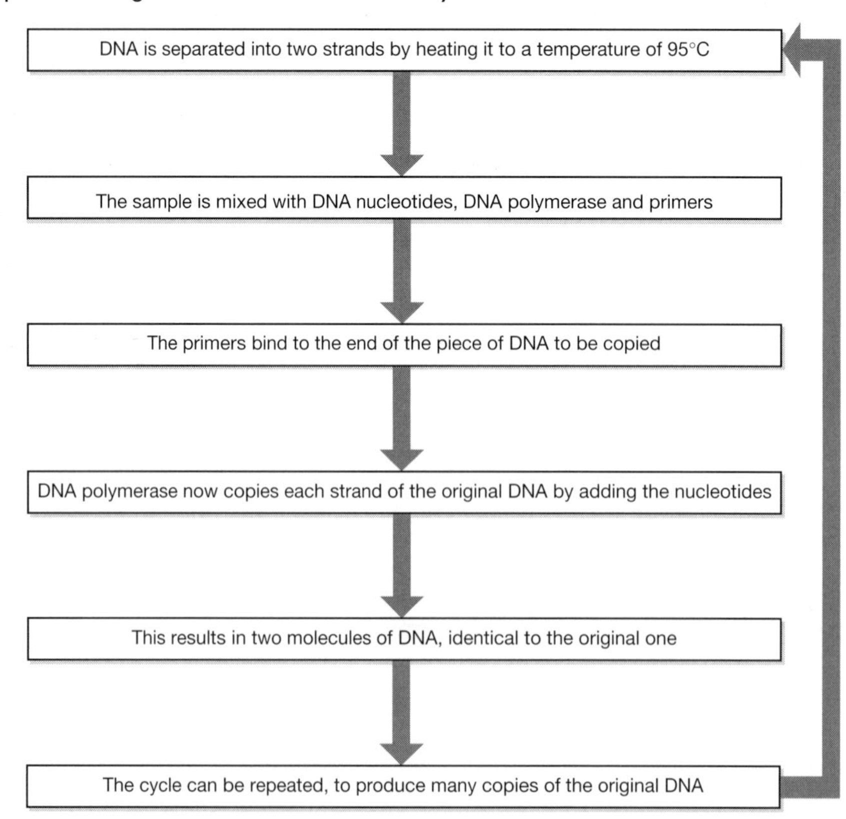

Questions to try

Examiner's hints

- Triplets of bases in DNA code for codons in mRNA, which have complementary anticodons on tRNA.
- During transcription mRNA is not made from DNA. The sugar and one of the bases in mRNA are different from those in DNA. DNA only acts as a template for the assembly of mRNA.

Q1

Cancer may be treated by chemotherapy. This involves using drugs which kill cancer cells but have no effect on normal healthy cells. Unfortunately, cancer cells develop from normal cells so the two types of cell are similar to each other. Trials have begun which involve adding a new gene to the normal cells in the body. This gene makes a protein which protects these healthy cells against the drug being used. The cancer cells don't produce this protein, so they are killed.

(a) Describe the features of a gene which enable it to code for a particular protein.

....

....

....

[4 marks]

(b) Describe how the new protein is made once the gene has been inserted into the cell.

....

....

....

....

....

[7 marks]

Examiner's hints

- Be sure you know which enzyme does what. A restriction endonuclease (restriction enzyme is enough) recognises specific sequences of base pairs and cuts between particular bases. DNA ligase joins the nucleotides within the backbone together to make recombinant DNA, not the nucleotides between the nitrogenous bases.
- Restriction enzymes that create a staggered end are more useful than those which produce a blunt end.

A restriction endonuclease cuts DNA at a particular base sequence. The restriction endonuclease *Bam* H1 recognises the sequence of six bases as shown in the diagram and cuts the DNA to form sticky ends. The arrows show where *Bam* H1 cuts the DNA.

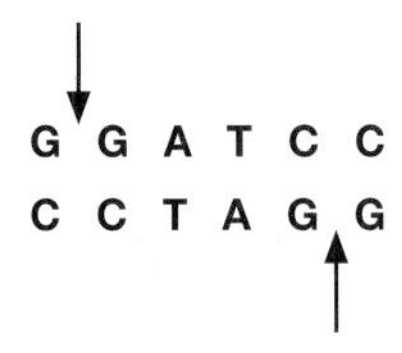

(a) Draw the sticky ends that are produced when *Bam* H1 has cut the DNA.

[1 mark]

(b) Describe how the two polynucleotide chains of DNA are normally held together.

....

....

....

[2 marks]

(c) The enzyme DNA ligase is used to join together pieces of DNA from different sources. Explain why the DNA to be joined together must be cut with the same restriction endonuclease before DNA ligase is used.

....

....

....

[2 marks]

[Total 5 marks]

Examiner's hints

- When converting from mRNA to DNA bases don't forget that DNA has no uracil and RNA has no thymine.
- Split the bases into triplets and read them three at a time. Not only does this relate to the way the code is read but it also makes your transcription errors less likely.
- Read the stem of the question carefully and relate it to the available marks. Underline key areas and be sure you spread your effort evenly.

Q3

Part of a particular messenger RNA (mRNA) molecule contains the sequence of nitrogenous bases opposite.

GGAGACUCCCCCAGUGUA

(a) Write the sequence of bases in the DNA strand from which this mRNA was derived.

[1 mark]

(b) Use the table, which shows the association between some amino acids and their appropriate messenger RNA (mRNA) bases, to determine which amino acids would be incorporated into the polypeptide obtained from this sequence.

mRNA bases	Amino acid	mRNA bases	Amino acid
UUC	phenylalanine	AUA	isoleucine
UCC	serine	ACG	threonine
UAC	tyrosine	AAU	aparagine
UGG	tryptophan	AGU	serine
CUA	leucine	GUA	valine
CCC	proline	GCC	alanine
CAA	glycine	GAC	asparagine
CGC	arginine	GGA	glycine

Amino acid sequence:

....

[1 mark]

(c) Most codons code for amino acids.

(i) What is meant by a codon?

....

[1 mark]

(ii) State another function that a codon may perform.

....

[1 mark]

(d) With particular reference to the role of transfer RNA and ribosomes, outline the stages involved in polypeptide synthesis after the formation of messenger RNA.

....

....

[4 marks]

Answers to Questions to try are on pp. 87 – 88.

[Total 8 marks]

Exam question and student's answer

The diagrams show part of the molecular structures of two polysaccharides. The hexagonal shapes represent hexose sugars.

Molecule A

Molecule B

Key: ------ Hydrogen bonds

(a) Give the name of molecule **A**.

Starch ✓ — 1/1

[1 mark]

(b) Give one difference between the hexose sugars in molecules A and B.

One is an alpha glucose and the other is a beta glucose. *which one?* — 0/1

[1 mark]

(c) Both polysaccharides contain hexose sugars joined by 1–4 glycosidic bonds.

(i) Explain, using an annotated diagram, how these bonds in molecule A are hydrolysed in the process of human digestion.

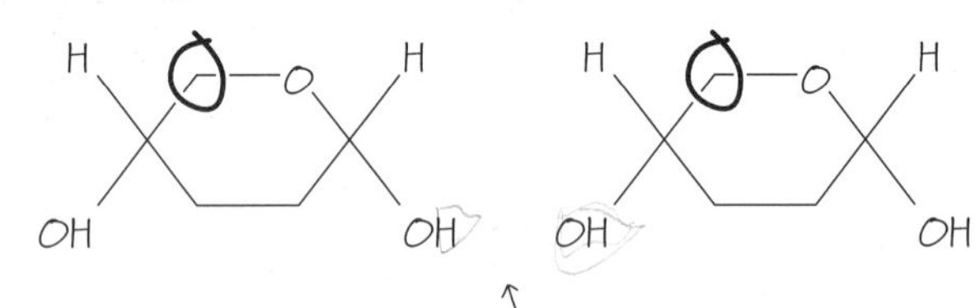

1/2

(ii) Using information in the diagram of molecule B, suggest one reason why it cannot be digested by humans.

Molecule B has hydrogen bonds while molecule A does not. These bonds cannot ✓ be broken down during the process of digestion. — 1/1

[3 marks]

[Total 5 marks] — 3/5

How to score full marks

Simple answers are best

There are a number of polysaccharides, all with the same basic branching structure as molecule A. They include amylopectin (one of the two polysaccharides that make up starch) and glycogen. **The obvious one is starch**, however, and is the one chosen by the student in part (a). **There are no extra marks for choosing the most obscure or rare examples so never try to be too clever.**

Don't be vague

Although the difference between the hexose sugars that make up these molecules is given correctly in part (b), **the student doesn't make it clear which polysaccharide is made of alpha glucose and which of beta glucose**. You should write 'Molecule A is made of alpha glucose while B is made of beta glucose'. This leaves the examiner in no doubt that you know the correct answer.

Copy what's there

If diagrams of molecules are given, always copy the whole thing. There is no point adding extra atoms to prove you know where they should be, but it is also important not to miss anything out. The most common omission made by candidates was made by this student in part (c)(i). He left out the line that represents the side chain. It cost him a mark! Make sure you copy corrcctly.

Use the information

In part (c) (ii) there are a number of reasons why cellulose, molecule B, cannot be digested by humans - for example, we don't produce the enzyme cellulase. However, that **cannot be seen in the diagram**. If you spot the most obvious difference between the two molecules (the presence of hydrogen bonds), as this student has done, you will get the mark. A correct, but really complex, answer is 'The shape of the molecule, caused by different glycosidic bonds between the beta glucose units, means that it will not fit into the active size of the enzyme that hydrolyses starch.' Can you fit that into two lines?

Don't make these mistakes ...

Key points to remember

Carbohydrates

Elements present:

- Carbon, hydrogen and oxygen

Basic building units/monomer:

- Monosaccharide
- Only hexose (6 carbon) sugars form polymers
- Other examples include pentose (5 carbon) sugars and triose (3 carbon) sugars

Polymer:

- Polysaccharide

Joining the units:

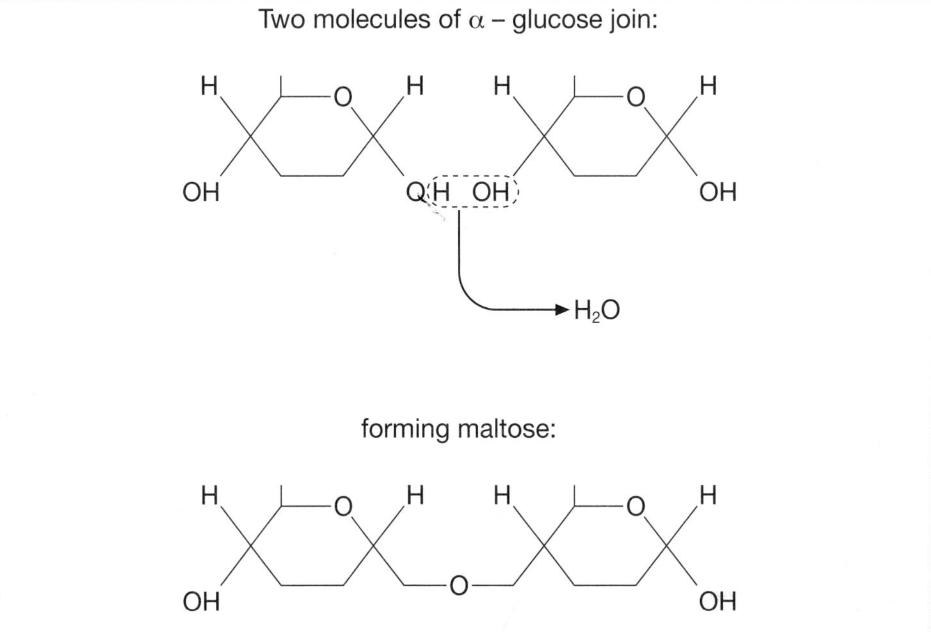

Two molecules of α – glucose join:

$\rightarrow H_2O$

forming maltose:

Type of bond joining monomer:

- Glycosidic

Further bonds:

- Hydrogen

Main function:

- **Starch** and **glycogen** are **storage** compounds. They can be broken down to form glucose.
- **Glucose** is important in **respiration**
- **Cellulose** is an important component of **plant cell walls**
- **Ribose** is a component of **nucleic acids**

Biochemical tests:

- **Reducing sugar test** – add **Benedict's reagent**, heat blue solution. An orange/red precipitate forms if reducing sugar present
- **Non-reducing test** – if no reducing sugar is present add **HCl**, boil; neutralise using **sodium hydrogen carbonate**; then perform Benedict's test. An orange/red precipitate will form
- **Starch test** – add **iodine solution**. Colour changes from brown to blue-black

Proteins

Elements present:

- Carbon, hydrogen, oxygen, nitrogen and often sulphur

Basic building units/monomer:

- Amino acids

Polymer:

- Polypeptide

Joining the units:

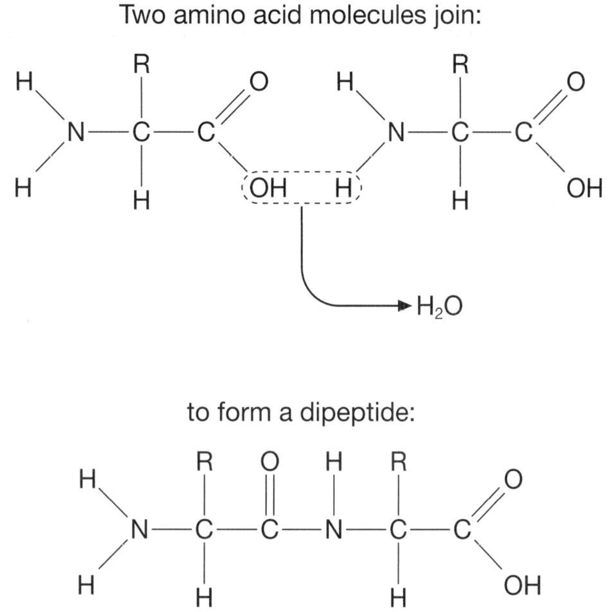

Two amino acid molecules join:

$\rightarrow H_2O$

to form a dipeptide:

Type of bond joining monomer:

- Peptide

Further bonds:

- Hydrogen
- Disulphide
- Ionic
- Covalent

Main function:

- Proteins have many functions. They act as **receptors**, they form **antibodies**, catalyse reactions as **enzymes** and are important **structural** substances in tissues such as muscles

Biochemical test:

- **Biuret test** – add **sodium hydroxide** followed by a few drops of **copper sulphate**. The colour changes from blue to mauve/purple

Key points to remember

Lipids

Elements present:

- Carbon, hydrogen and oxygen

Basic building units:

- Fatty acids and glycerol

Macromolecule:

- Lipids – triglyceride

Joining the units:

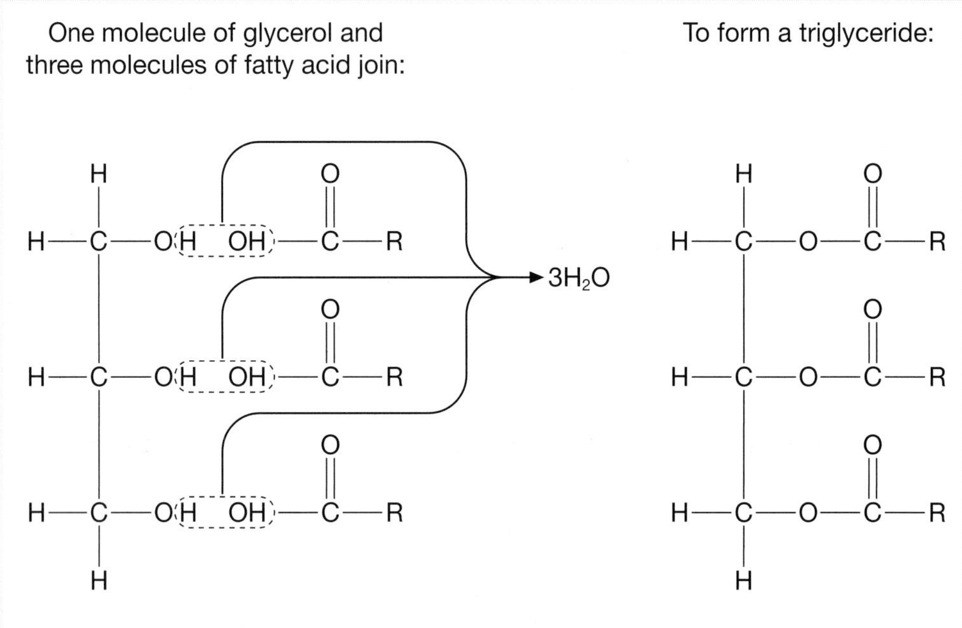

One molecule of glycerol and three molecules of fatty acid join:

$\rightarrow 3H_2O$

To form a triglyceride:

Type of bond:

- Ester

Further bonds:

- None

Main function:

- Lipids are important **storage** compounds and can be used in **respiration**
- Modified lipids with one fatty acid replaced by a phosphate group form **phospholipids** – components of **membranes**

Biochemical test:

- **Emulsion test** – add **ethanol**; shake; pour off into water. A white emulsion forms on top of the water

Questions to try

Examiner's hints

- If the question asks for a comparison, as part (a) does, focus on the correct side of that comparison. In this question to describe proteins or carbohydrates and then to write that lipids are different would not help.
- Never do too much. Part (c) asks you to show how a single fatty acid joins with glycerol. The process is the same for the others. You will be wasting valuable time – and will get no extra marks – if you repeat yourself.

(a) Give **two** ways in which lipid molecules differ from other macromolecules such as proteins and polysaccharides.

1 ....

2 ....

[2 marks]

(b) Describe **one** test that would enable you to confirm the presence of lipid in a maize grain.

....

....

[1 mark]

(c) The diagram shows the structure of a glycerol molecule and of a fatty acid molecule.

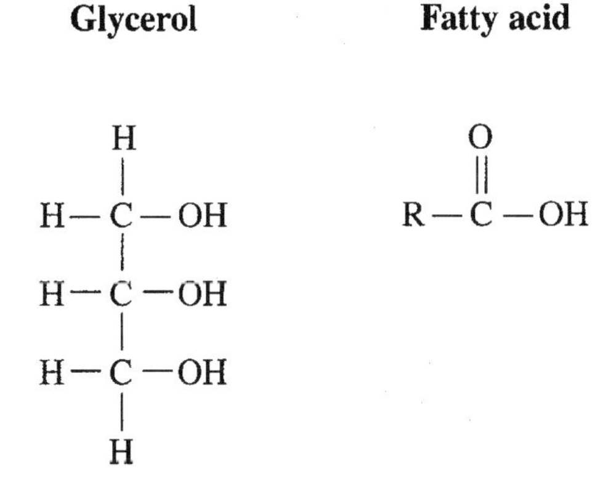

A glycerol molecule can join with one, two or three fatty acid molecules to produce a monoglyceride, a diglyceride or a triglyceride. With the aid of a similar diagram show how the molecules shown above may join in a condensation reaction to form a monoglyceride.

[2 marks]

Examiner's hints

- The same question could be asked at GCSE level, at 'A' level and even at degree level. Be sure you give an 'A' level answer.
- Explain means 'give a reason for ...' *not* 'write all you know about ...'.

Q2

Sucrose is a disaccharide. The diagram shows the structure of a molecule of sucrose.

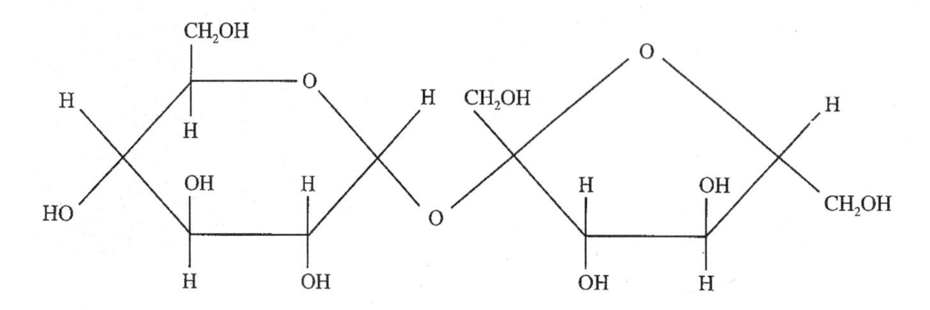

(a) (i) Use the diagram to explain why sucrose is classified as a carbohydrate.

....

....

[1 mark]

(ii) Explain why sucrose will produce a positive result with Benedict's test only after it has been boiled with a dilute acid.

....

....

....

[2 marks]

(b) Sucrose is sweet-tasting. The receptor molecules in the taste buds on the tongue are proteins. They detect sweet-tasting substances only if they have dissolved in the saliva.

Explain how proteins are suited for their roles as receptor molecules.

....

....

....

[3 marks]

Examiner's hints

- Questions relating to biological molecules frequently follow the same pattern. The following 'bits' of questions look for the same principles.
  - To **break** two units apart you add a water molecule: a **hydrolysis**.
  - To **join** two units together you remove a water molecule: a **condensation**
- When questions ask you to show with similar diagrams how two units join - monosaccharides, amino acids, or fatty acid and glycerol – always **copy the whole molecule**, always **show water being removed**.
- Remember the names of the bonds formed by each group of molecules.
  - Two monosaccharides form a glycosidic bond
  - Two amino acids form a peptide bond
  - Fatty acids and glycerol form ester bonds
- Chemical tests are important. Try to remember what you add, what colour it is or what it looks like at the beginning, what you do, what colour it is or what it looks like at the end.

(a) The figure shows the structure of an amino acid molecule.

R
H, H – N – C(R)(H) – C(=O)OH

(i) By means of a similar diagram, show how two amino acid molecules may be joined together with a peptide bond.

[2 marks]

(ii) What is the name given to the type of chemical reaction in which two amino acids are joined together?

....

[1 mark]

(b) The diagram shows the structure of a molecule of α-glucose.

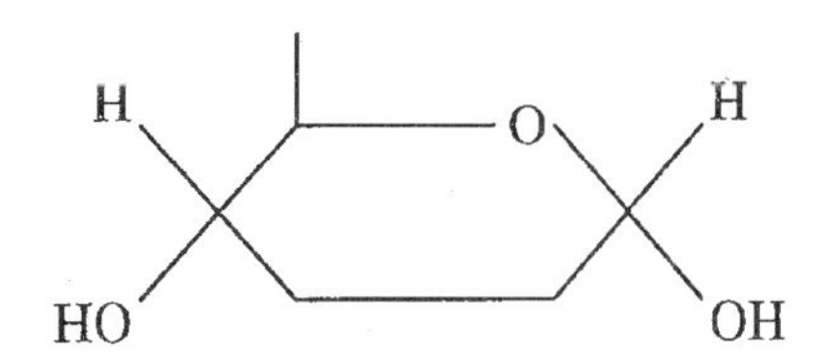

(i) With the aid of similar diagrams show how two α-glucose molecules may be linked by means of a glycosidic bond.

[2 marks]

(ii) Describe a biochemical test that would allow you to distinguish between solutions of glucose and sucrose.

....

....

[2 marks]

Answers to Questions to try are on pp. 89 – 91.

Exam question and student's answer

(a) Explain why it is possible to see cell structure in more detail with an electron microscope than with an optical microscope.

Electron microscopes magnify things more and so you will be able to see more information. Things will look bigger and you will be able to see them clearly.

[2 marks]

(b) The drawing shows a group of prokaryotic cells.

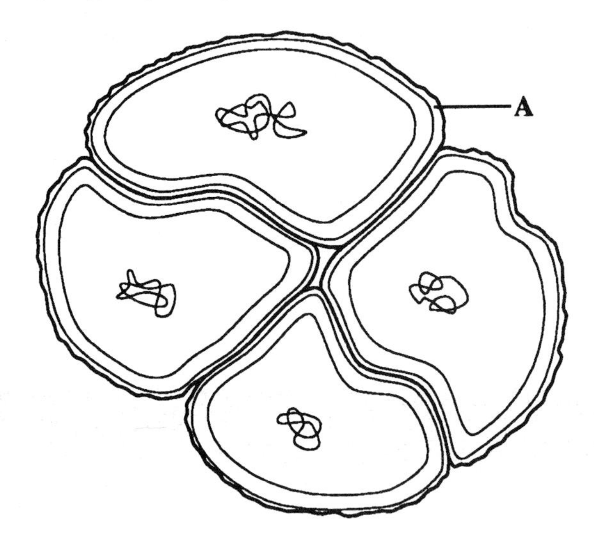

(i) Identify feature **A**.

Feature A is the capsule around the outside of the prokaryotic cell ✓

(ii) Give **two** pieces of evidence from the drawing to support the fact that these are prokaryotic cells.

1 The cells have a capsule around the outside

2 The cells have no nucleus ✓

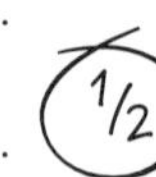

[3 marks]

[Total 5 marks] 2/5

How to score full marks

Get your terms right

The advantage of an electron microscope is *not* that it makes things bigger (i.e. magnifies things), as the student suggested in part (a). To gain the first mark you need to say that **it has a greater *resolution* at the same magnification, so you can see things more clearly**.

Watch the mark allocation

The student didn't write anything that could gain the second mark in part (a) because the second sentence simply restates the first. Why does an electron microscope have a better resolution? **Because electrons have shorter wavelength** is the correct answer, which would gain you the second mark.

Two marks for a single idea – I don't think so!

In (b) (ii), using a piece of information from another part of the question is unlikely to get you credit. **'There are no membrane-enclosed organelles' or 'no named organelles (like mitochondria) are visible' would be much better answers.**

Don't generalise

In part (b) (ii) an answer such as 'prokaryotic cells do not have membrane-enclosed organelles' is correct, but if you write only that you may only get a single mark. **If you name organelles** – mitochondria, nucleus, chloroplast, etc. – **you could get a mark for each.**

Don't make these mistakes ...

Magnifying things does not make them clearer. You must refer to resolution. Electron microscopes use a form of radiation, which has a shorter wavelength and therefore is able to distinguish between two objects, which are very close together. Light, with a longer wavelength, will make these appear as a single point.

A bacterium is an example of an organism with a prokaryotic cell structure. **They have few organelles – and none that are membrane enclosed**, such as mitochondria or chloroplasts. However, they do contain ribosomes.

Key points to remember

Cell types

All living organisms are made up of cells, but there are two main types.

- **Prokaryotic** cells, such as bacterial cells
- **Eukaryotic** cells, found in all other organisms including plants, animals, fungi and protoctista.

The table shows the main differences between these two types of cell.

Prokaryotic cell	Eukaryotic cell
Small cells 5 μm	Larger cells up to 50 μm
No nucleus	Nucleus enclosed by membrane
DNA is a circular strand	DNA is a linear strand
Few organelles present – none with a membrane surrounding it	Many membrane-bound organelles, e.g. mitochondria, chloroplasts
Ribosomes are small and free in the cytoplasm	Ribosomes are large and often associated with the membrane of the rough endoplasmic reticulum

Organelles

Eukaryotic cells contain many organelles. It is important to know which are in plant cells and which are in animal cells and their functions.

Organelle	Present in plant cells	Present in animal cells	Function
Cell wall	✓		Strength. Resists the pressure created when water enters
Plasma membrane	✓	✓	Selectively controls the movement of substances into and out of the cell
Nucleus	✓	✓	Contains DNA, which holds the genetic information
Mitochondria	✓	✓	Produce large amounts of ATP by aerobic respiration
Chloroplast	✓		Photosynthesis
Rough endoplasmic reticulum (ribosomes)	✓	✓	Protein synthesis
Smooth endoplasmic reticulum	✓	✓	Synthesis of lipids
Golgi apparatus	✓	✓	Joining carbohydrates to proteins to form glycoproteins and the secretion of carbohydrates to form new cell walls

Microscopes

- These allow you to see small objects clearly.
- There are two main types, **light microscopes** and **electron microscopes**.
- The **resolution** of the electron microscope is much better due to the very small wavelength of the electrons. This means that if both magnified an image to the same level the object would be seen more clearly using an electron microscope.

Light microscope	Electron microscope
Uses light	Uses a beam of electrons
Specimen may be living or dead	Specimen must be dead
Magnification up to 1500 times but most at 400 times	Magnification up to 500 000 times
Resolution 0.2 µm	Resolution 0.001 µm

Electron microscopes

There are two main types of electron microscope.

- **Transmission:** A beam of electrons passes through the object and shows the internal structures in detail.
- **Scanning:** A beam of electrons is reflected off the surface. This produces a three-dimensional view of the surface.

Cell fractionation

This is a way of separating organelles.

- The tissue or cells are broken up in a solution, either mechanically or manually.
- The solution must be:

  a buffer to keep the pH constant – if the pH changes the enzymes of the organelles could be destroyed

  cold – digestive enzymes within the cell would destroy the organelles at higher temperatures

  at the same water potential, so there is no net movement of water into or out of the organelle – this prevents it from bursting or shrivelling.
- The mixture is filtered to remove any large unwanted debris.
- The filtrate is centrifuged at low speed. This makes large organelles such as the nuclei fall to the bottom of the tube and produce a pellet. The remaining liquid (the supernatant) will contain all the lighter organelles.
- The supernatant is centrifuged at a higher speed to make the larger organelles left suspended in the liquid form a pellet.
- This can be repeated at higher speeds and for longer times to remove successively smaller organelles.

Conversions

Biologists tend to measure small things, so you must be confident in using **millimetres** (mm), **micrometres** (µm) and **nanometres** (nm). They have an easy to remember relationship and conversion is really straightforward.

- mm = 10^{-3} metres
- µm = 10^{-6} metres
- nm = 10^{-9} metres

Don't panic!
10^3 means 10 × 10 × 10 = 1000, so 3 noughts.
10^{-3} means 3 noughts after the decimal point – 0.00010 or $^1/_{10\,000}$ th.

To convert one to the other

- First decide whether the conversion is going to produce a larger or smaller figure.
  If you need to convert 50 mm to µm, remember:

  1 µm is smaller than 1 mm so there will be a greater number of µm

  But If you need to convert 50 nm to µm, remember that 1 µm is bigger than 1 nm so there will be a smaller number of µm
- Secondly move the decimal point three places.

Key points to remember

Calculating size

- Always measure in mm
- Convert to real size using the magnification or the scale given
- Use the most appropriate unit
- Convert using the method described on page 59

If you are given the magnification

You will often be required either to calculate the magnification of an object or to work out its actual size using a magnification given in the question. Basically there are three things you have to know:

- The magnification (M)
- The size of the object that you can physically measure, the observed size (O)
- The actual size of the object (A)

These can be put into a simple equation $M = \frac{O}{A}$

You can use this equation depending on the information you are given and what you are asked to calculate. For example:

- The magnification of a mitochondrion (M) was 1500 times, the measurement you made of it with your ruler (O) was 15 mm. How do you work out the actual size?;

Rearrange the equation above to give you the actual size: $A = \frac{O}{M}$

Put in the figures you have got: $A = \frac{15}{1500} = 0.01 \text{ mm}$

So the actual size of the mitochondrion is 0.01 mm (don't forget the units) or 10 μm.

If you are given the scale

How wide is this cell?

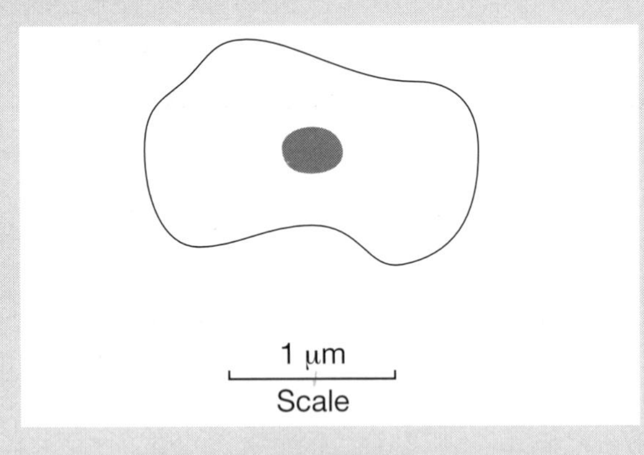

The diagram shows that the cell is 40 mm wide and the scale shows that 20 mm represents 1 μm

As 40 mm represents 2 μm the cell is really 2 μm wide.

The pattern of what to do is easy to follow.

If 20 mm represents 1 μm, a distance of 40 mm will represent

$$40 \times \frac{1}{20} \ \mu\text{m} = 2 \ \mu\text{m}$$

If 10 mm represents 6 μm, a distance of 14 mm will represent

$$14 \times \frac{6}{10} \ \mu\text{m} = 8.4 \ \mu\text{m}$$

Questions to try

Examiner's hints

- Look at the distance represented by the scale.
- Look at the relationship between that distance and the object. Is it bigger or smaller?
- Be sure your answer matches this rough estimate.
- If units are not given be sure to add them.

Q1

(a) The drawing has been made from an electronmicrograph and shows part of the cell walls of two plant cells.

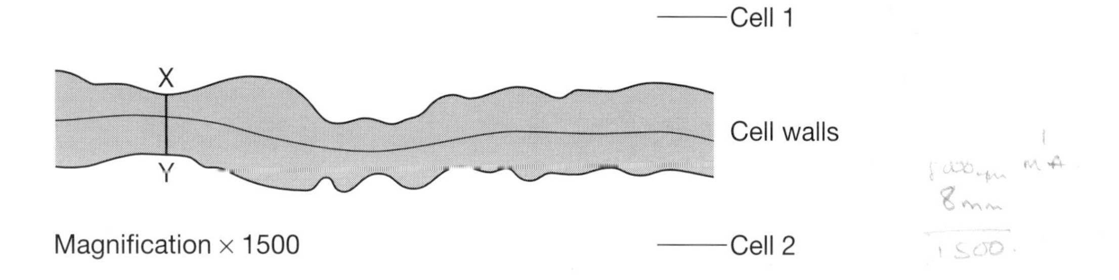

Calculate the actual thickness of the walls in micrometres (μm) between points **X** and **Y**. Show your working.

....

[2 marks]

(b) Plant cell walls contain a number of different chemical compounds including cellulose and glycoproteins.

Explain **one** way in which cellulose molecules are adapted to their function in a plant cell wall.

....

[1 mark]

(c) In which part of the cell are:

(i) proteins synthesised;

....

[1 mark]

(ii) glycoprotein molecules assembled?

....

[1 mark]

Examiner's hints

- Think of organelles in terms of their density or mass, never in terms of their size. Structures can be big but light or small but heavy. Certainly there would also be a range of size dependent on the maturity of the organelle but with the same density they should all be found in the same area.
- The process of respiration occurs in the cytoplasm and mitochondria. Only the aerobic stages happen in the mitochondria.

Cell organelles can be separated by centrifuging a cell extract in a sucrose density gradient. The organelles settle at the level in the sucrose solution which has the same density as their own.

Some animal cells were broken open and the cell extract centrifuged in a sucrose density gradient. Three distinct fractions were obtained, **A**, **B** and **C,** as shown in the diagram.

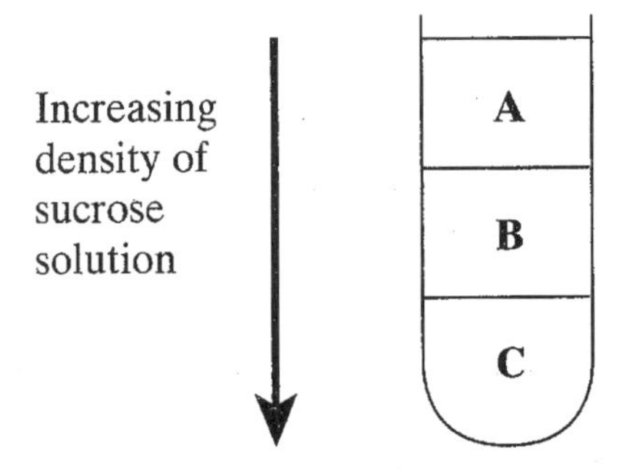

One fraction contained nuclei, one contained ribosomes and a third contained mitochondria.

Complete the table by identifying the organelle in each fraction and describing **one** function of each organelle.

Fraction	**Organelles**	**Function**
A		
B		
C		

[4 marks]

Answers to Questions to try are on p. 92.

Exam question and student's answer

The diagram shows a vertical section through a human heart. The arrows represent the direction of movement of the electrical activity which starts muscle contraction.

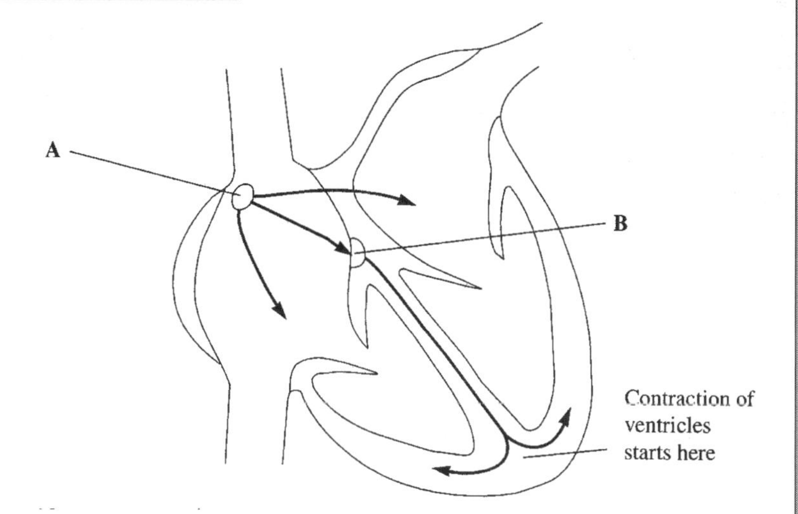

(a) Name structure **A**.

Structure A is the SAN

[1 mark]

(b) Explain why each of the following is important in the pumping of blood through the heart.

(i) There is a slight delay in the passage of electrical activity that takes place at point B.

Structure B is the AVN and it has to be stimulated which takes time.

[1 mark]

(ii) The contraction of the ventricles starts at the base.

Blood leaves the heart in the pulmonary artery and the dorsal aorta at the top of the heart, so blood has to be pushed upward *

[1 mark]

(c) Describe how stimulation of the cardiovascular centre in the medulla may result in an increase in heart rate.

The cardiovascular centre sends a message along a nerve to the SAN which makes it work faster.

[2 marks]

* so the heart beat has to start from the bottom to push the blood in that direction into the arteries.

(d) Arteries may become blocked by the formation of fatty material on the walls. An operation called balloon angioplasty may be used to correct this. The procedure is shown in the diagram.

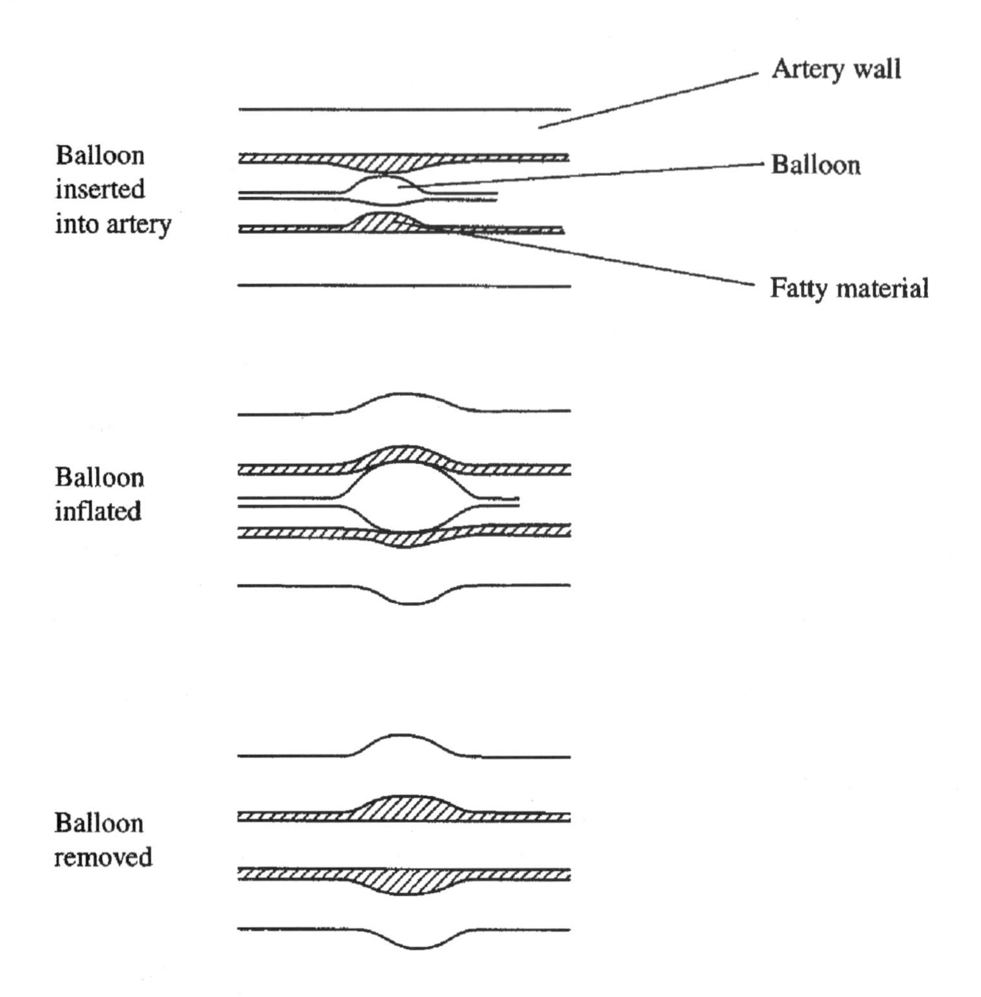

(i) Suggest why the artery wall 'bounces back' when the balloon is removed.

The wall of the artery is made of muscle and this contracts to push the blood through.

0/1

(ii) Explain why the ability of the artery wall to bounce back is important in a normal, healthy artery.

Blood comes out of the heart in pulses ✓ and the muscle contraction helps to even ✓ out the jumps in pressure.

2/2

[3 marks]

[Total 8 marks] 5/8

How to score full marks

Don't take risks

'SAN', the student's answer for part (a), is the abbreviation for the **sino-atrial node** or the pacemaker. Although in this case there can be no other meaning for these letters and it gained the mark, some of the abbreviations students use are their own (or their teacher's) attempt to make a name easy to remember. **In an exam, use the correct term whenever possible**.

Know the function

Although the idea the student gave in part (b) (i) is reasonable **it does not answer the question**. The correct answer is that **the delay allows complete contraction of the atrium so that all the blood is in the ventricle before the ventricle begins to contract**.

Watch the space and the mark allocation

There is only a single mark for part (b) (ii) but the candidate has written loads!! **The only thing that was needed was to recognise that blood has to flow upwards to leave the heart**. He got his mark right at the beginning of his answer but gave a lot of extra material and restated the question. This wastes time and the chance of picking up marks elsewhere.

Use the right language

In part (c) the student actually offered three possible ideas for the two marks. **In two of the cases he fell short of what was expected**. Nerves do not carry messages – they carry impulses. Along a nerve – what nerve? The involvement of the sympathetic nervous system was needed for this mark.

The SAN 'works faster' was allowed as a borderline point (see the tick plus on the answer) – **increases in activity** would be a better way of expressing this idea.

Generous examiners

A mistake in one part of question (d) (i) (**the artery contains elastin**, a protein which recoils when stretched – this allows the recovery of the artery, **not muscle** as the student suggested) was ignored by the examiner in the second part. The student correctly described the purpose of the recoil and was awarded the marks.

Don't make these mistakes ...

All structures have a role or 'job' to do. **There is no point remembering the structure without remembering what it does.** Actually link the two together and they both become easier to remember.

There are a number of acceptable abbreviations used in biology. ATP, DNA, NAD are examples. No one wants students to waste time remembering long complicated names for the sake of the exercise. However, **you should remember the names of structures** like the sino-atrial node.

Some questions come right from your syllabus – part (d) of the example question here, for instance. Don't ignore that when you revise – **use your syllabus as a starting point and be sure that you can do what it asks.**

Key points to remember

Blood vessels

Blood leaves the heart in **arteries**
↓
It then flows through **arterioles**
↓
To **capillaries**
↓
This is where exchange of substances with cells/tissues takes place
↓
It is collected in **venules**
↓
And flows back to heart in **veins**

Arteries and veins have similar structures but capillaries are very different. The table shows some of the similarities and differences between arteries, veins and capillaries

	Arteries	**Veins**	**Capillaries**
Walls	Three layers: *inner* – smooth endothelium cells lining the lumen and some muscle; *middle* – thick and muscular with many elastic fibres; *outer* – layer of collagen	Three layers: *inner* – smooth endothelium cells lining the lumen; *middle* – almost non-existent with very little muscle; *outer* – layer of collagen	One layer: only smooth endothelium lining
Size of lumen	Relatively small	Relatively large	Same diameter as red blood cell
Blood pressure	High	Low	Medium
Valves	Absent	Present	Absent
Function	Transport	Transport	Exchange

The heart

The heart is effectively two pumps, which are coordinated and beat together.

- The right side is a low-pressure system pumping blood a short distance to the lungs.
- The left side is a high-pressure system pumping blood long distances all over the body.

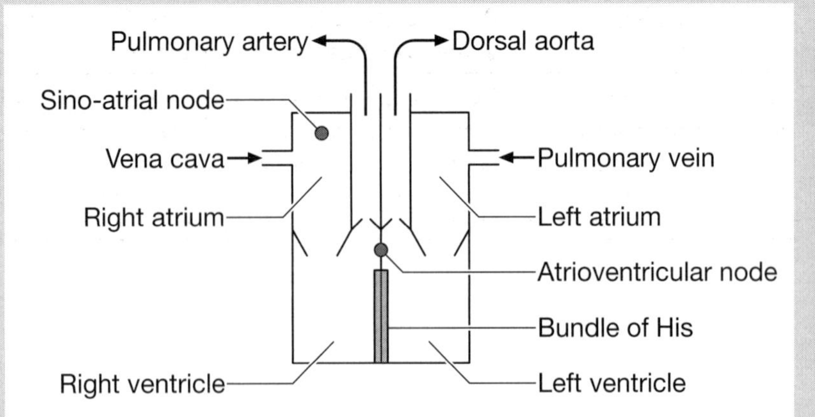

Pressure and volume

- Blood always moves from an area of high pressure to an area of low pressure.
- As the heart chambers contract the volume is reduced and the pressure is increased.
- As the heart chambers relax the volume is increased and the pressure in decreased.

Coordinating the heartbeat

- Impulses originate in the sino-atrial node (SAN).
- The impulses spread through the walls of the atria, causing the muscle to contract and moving blood into the ventricles.
- The impulse arrives at the atrioventricular node (AVN).
- The impulses can only travel to the ventricles from the AVN.
- The signal is delayed by the AVN, allowing time for the atria to complete their contraction and thus for all the blood to move into the ventricles.
- The AVN passes an impulse through the bundle of His to the base of the heart.
- The impulse spreads through the heart from the base upward, causing the muscle to contract as it goes and moving the blood into the arteries.
- The rate at which the SAN initiates impulses – and therefore the rate at which the heart beats – can be **increased** by hormones (for example adrenaline) or by the sympathetic nervous system; **decreased** by the parasympathetic nervous system.

The cardiac cycle

This is a continuous cycle but can be divided up into a number of stages.

1 Atrial systole:

- The walls of the atria **contract**.
- This **reduces the volume** of the atria, and therefore **increases the pressure**.
- Blood is forced **through** the **atrioventricular valves**
- into the ventricles.

2 Ventricular systole:

- The walls of the ventricles **contract**.
- This **reduces the volume** of the ventricles, and therefore **increases the pressure**.
- Blood is forced **against** the atrioventricular valves – closing them – and **through** the **semilunar valves**
- into the arteries.

3 Ventricular diastole:

- The walls of the ventricles **relax**.
- This **increases the volume**,
- and therefore **decreases the pressure**.
- Blood starts to flow into the ventricle from the atrium again.

Conditions that cause the heart valves to open and close

Atrioventricular valve *open*	Atrioventricular valve *closed*	Semilunar valve *open*	Semilunar valve *closed*
Pressure is higher in atrium than ventricle. Blood will flow from atrium to ventricle	Pressure is higher in ventricle than atrium, preventing blood flow into atrium	Pressure is higher in ventricle than artery. Blood will flow from ventricle into artery.	Pressure is higher in artery than in ventricle, preventing back-flow of blood into ventricle

Questions to try

Examiner's hints

- Read the stem of the question carefully, especially if there is a long passage. Note exactly what the graph shows – you **will** be expected to use the information.
- The heart is always drawn as if the animal involved is lying on its back and you are looking down on it. So your left is the right-hand side of the heart.
- If you are asked to fill in a table read the instructions carefully. Do you need to use a tick or a cross or both? If you make a mistake don't attempt to alter the tick to a cross – score it out and start again.

Q1

A small tube called a catheter can be inserted into the blood system through a vein. It can be threaded through the vein and into and through the heart until its tip is in the pulmonary artery. A tiny balloon at the tip can then be used to measure the pressure changes in the pulmonary artery. The diagram shows a section through the heart with the catheter in place. The graph shows the pressure changes recorded in the pulmonary artery.

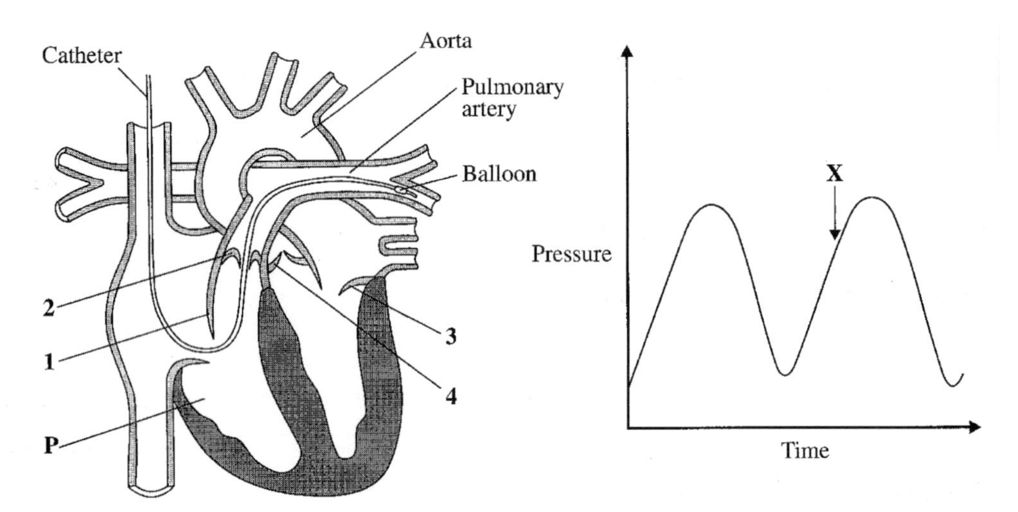

(a) Name the chamber of the heart labelled **P**.

....

[1 mark]

(b) Complete the table by placing ticks in the appropriate boxes to show which of values **1** to **4** will be open and which closed at time **X** on the graph.

[2 marks]

Valve	Open	Closed
1		
2		
3		
4		

(c) Sketch a curve on the graph to show the pressure changes that you would expect if the pressure in the aorta were measured at the same time.

[2 marks]

[Total 5 marks]

Examiner's hints

- The term 'cardiac cycle' describes the changes that occur in the heart, causing the movement of blood. If you are asked to describe what is happening in the heart during any phase of this cycle always write which valve is open and which is shut and which chamber is contracting and which is relaxing.
- Even if there is only one mark available, the term 'explain' means you have to give a reason for something. That is sometimes difficult unless you have described the situation first.

Q2

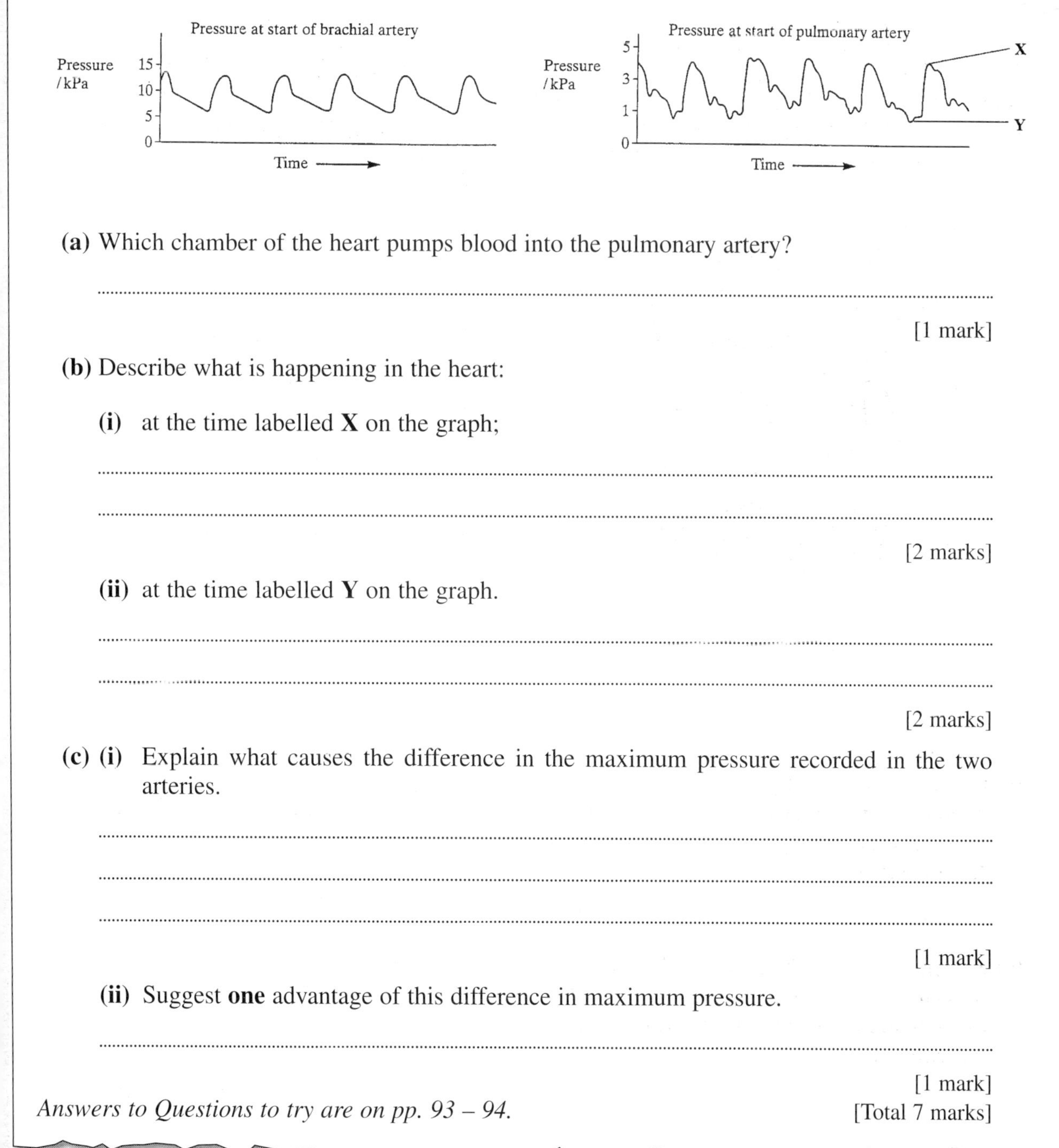

The graphs show how the pressure of blood in two arteries in a healthy person varies with time. The brachial artery supplies blood to the muscles of the arm.

(a) Which chamber of the heart pumps blood into the pulmonary artery?

....

[1 mark]

(b) Describe what is happening in the heart:

(i) at the time labelled **X** on the graph;

....

....

[2 marks]

(ii) at the time labelled **Y** on the graph.

....

....

[2 marks]

(c) **(i)** Explain what causes the difference in the maximum pressure recorded in the two arteries.

....

....

....

[1 mark]

(ii) Suggest **one** advantage of this difference in maximum pressure.

....

[1 mark]

Answers to Questions to try are on pp. 93 – 94.

[Total 7 marks]

Exam question and student's answer

(a) The graph shows how ventilation rate changes as a result of increasing the concentration of carbon dioxide in inspired air.

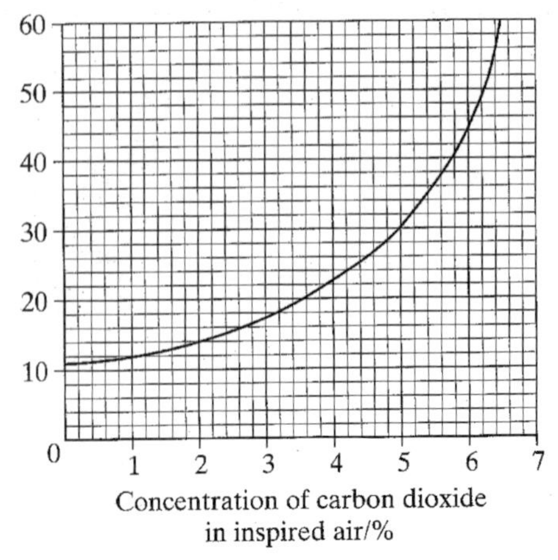

(i) Use the graph to calculate the percentage increase in ventilation rate when the concentration of carbon dioxide in inspired air rises from 0% to 5%.

$\frac{30 - 11}{11} \times 100 = 172.7272\%$ 172%

0/1

(ii) Explain how a rise in carbon dioxide concentration brings about the changes in breathing rate.

Increased levels of carbon dioxide in the blood ✓ are detected by the hypothalamus that sends impulses to the lungs to increase ✓ the rate of ventilation.

2/3

(iii) In mouth-to-mouth resuscitation a person breathes into the mouth of someone who has stopped breathing. This is more effective than other methods of resuscitation that involve altering the volume of the thorax by pressing on the chest wall.

Use the graph to suggest why mouth-to-mouth resuscitation is an effective way of restarting breathing.

The graph shows that increased levels of carbon dioxide such as you would expect in exhaled air would increase ✓ the rate of breathing. So using this method may encourage someone who had stopped breathing to start again, whereas pressing on the chest wall will just push normal air into his lungs.

1/1

[5 marks]

(b) The diagram shows an alveolus and its associated blood. Alveoli are adapted for efficient gas exchange by having very thin walls.

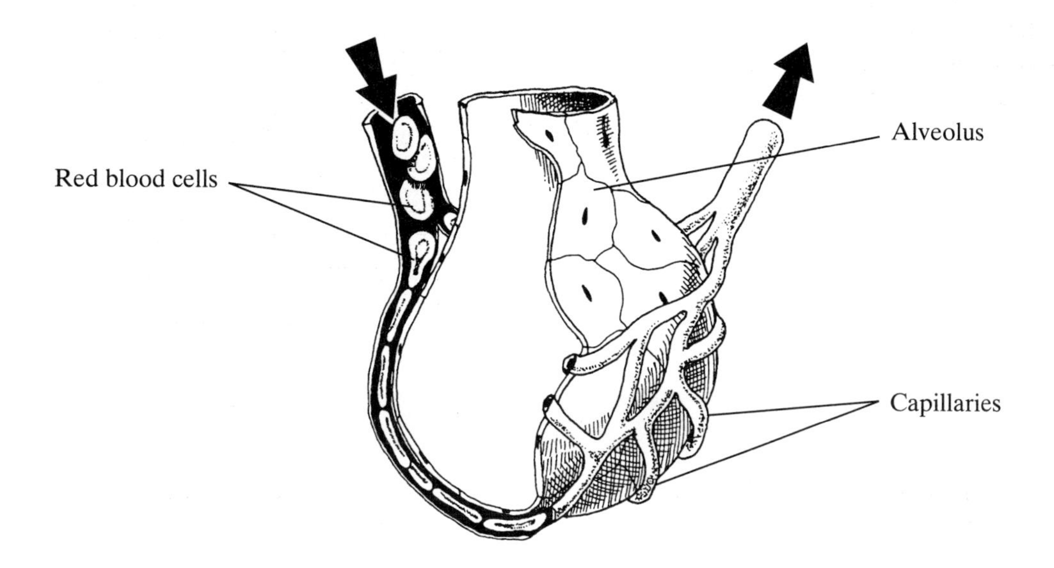

Suggest **one** other way, visible in the diagram, that these structures are adapted for efficient gas exchange.

Because there are so many capillaries there is a large surface area between the blood and the air in the alveoli.

1/1

[1 mark]

[Total 6 marks] 4/6

How to score full marks

Simple maths

You must be comfortable with some basic maths. **Percentage increases and decreases are very common questions**. Divide the change in value by the original value and multiply by 100.

In part (a) (i) the student got it right up to this stage but then rounded 172.7272 (the reading she got on her calculator) to 172, when it is almost 173.

The stem of the question asked for the **percentage increase** but if it had simply asked for the **percentage change** remember to write it had increased by 173%

Learn the links

There are a number of receptors in the body that detect changes in the physical and chemical nature of the blood. Increased and decreased levels of **glucose** are detected by the **pancreas**. Changes in **blood temperature** are detected by the **hypothalamus.**

How to score full marks

Carbon dioxide and **low pH** levels are detected by the **carotid bodies** – not by the hypothalamus as the student suggested in her answer to part (a) (ii).

An area of the brain called the **medulla** receives nervous impulses from the carotid bodies and coordinates the muscular contractions, which increase ventilation rate.

Look at the data

Part (b) asks you to give a feature 'visible in the diagram'. So, although there are a number of ways in which the lung is adapted for effective gas exchange, only the one given by the student and the thin diffusion pathway offered by the alveoli (epithelium) cells are **visible features**.

Don't make these mistakes ...

If you have to use your calculator remember two things:

1. **Make an estimate of the value you expect** before you start.

   In part (a) (i) the ventilation rate went from 11 up to 30, so it increased. If it had increased by 100% (i.e. if it had doubled) it would have reached 22 dm^3 per minute. If it had tripled (i.e. increased by 200%) it would have reached 33 dm^3. So an estimate is between 100 and 200%

2. Despite what your calculator is capable of doing, **offer a sensible value as an answer**.

   If you are capable of reading the scale on the graph to the nearest whole number – which is the case in part (a) (i) – and you are asked to calculate a mean value (the average), it would be silly to offer that to six decimal places. It's more sensible to give the percentage to the nearest whole figure – and so round up.

If only a single mark is allocated but three lines have been left for your answer, **look for the idea that will get you the mark and then stop**. Adding extra, rephrasing the question, or writing what you have just written in another way will not get you any extra marks.

Key points to remember

Size and surface area: volume ratio

The table compares the ratio between surface area and volume of cubes of different length sides.

Length of side	Surface area	Volume	Surface area : volume ratio
1	6	1	6 : 1
2	24	8	3 : 1
3	54	27	2 : 1

- Efficiency of diffusion depends on surface area, so very small organisms can meet all their gas exchange needs by diffusion through their surface.
- Larger organisms need specialised gas exchange systems that increase the surface area over which diffusion can take place.

Factors determining the rate of diffusion

- These factors can be summarised as Fick's law:

$$\text{Rate of diffusion} \propto \frac{\text{Surface area} \times \text{difference in concentration}}{\text{Thickness of gas exchange surface}}$$

Therefore the most effective gas exchange system must:

- provide as large a surface area as possible
- maintain as large a concentration gradient as possible
- provide as short a distance as possible for the gas to travel.

Gas exchange in humans

The lungs of a mammal achieve a very efficient gas exchange system in the following ways.

- Large surface area for gas exchange to occur, which is achieved by: **1** possessing many alveoli; **2** the lung capillaries having a very large total surface area.
- Maintenance of a large concentration gradient, which is achieved by: **1** ventilation – fresh oxygen is supplied to the lungs every time you breathe in, therefore keeping a high concentration of oxygen in the lungs; **2** circulation – capillaries are continuously taking away oxygenated blood and replacing it with deoxygenated blood.
- Thin gas exchange surface, which is achieved by: **1** there being only two cell layers for oxygen to pass through – the alveolar epithelium and the capillary endothelium; **2** both of these cell types being squamous epithelial cells, which means that they are very thin.

Ventilation

Movement of air is always down a pressure gradient, from an area of high pressure to an area of lower pressure.

Breathing in	Breathing out
Intercostal muscles and diaphragm muscles contract	Intercostal muscles and diaphragm muscles relax
Rib cage moves upwards and outwards and the diaphragm flattens	Rib cage moves downwards and inwards and the diaphragm returns to its dome shape
Volume of thorax increases	Volume of thorax decreases
Pressure decreases	Pressure increases
Air moves into lung	Air moves out of lung

Key points to remember

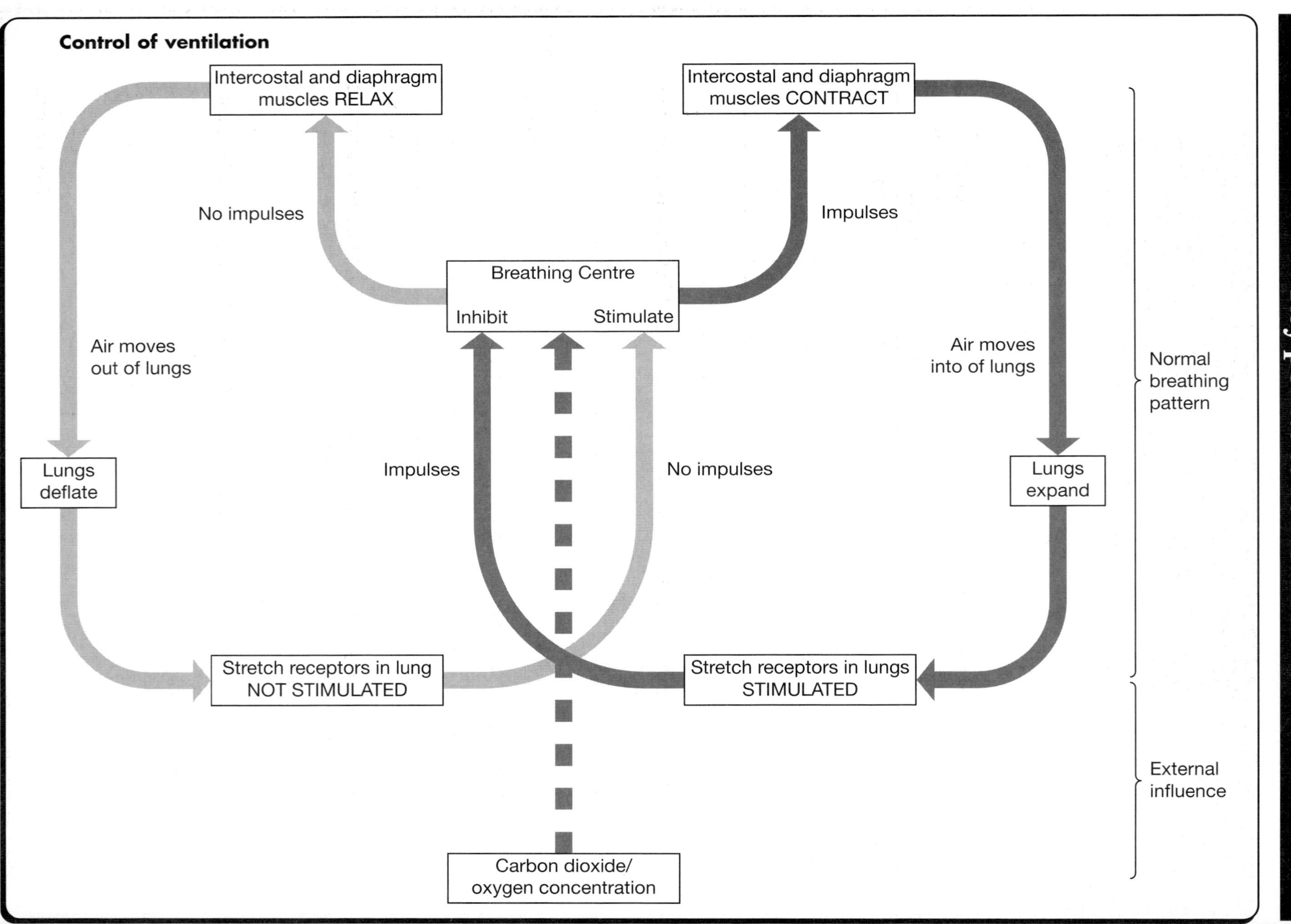

Questions to try

Examiner's hints

- If data in a table shows comparative amounts or percentages, then the amount of one thing is influenced by the addition or removal of something else.
- Be aware of the properties of the common gases of the air and what happens to these gases when they get into the lung.

Q1

The table shows the percentage of various gases in atmospheric air, exhaled air and in air samples collected from the alveoli and the trachea of a healthy human.

Gas	Atmospheric air (inhaled air)	Exhaled air	Sample of air collected from alveoli	Sample of air collected from trachea
Oxygen	20.9	15.7	13.7	19.6
Carbon dioxide	0.04	3.6	5.2	1.1
Water vapour	0.5	6.2	6.2	6.2
Nitrogen	78.5	74.4	74.9	73.1

(a) Use the table to suggest an explanation for the difference between:

(i) the percentage of carbon dioxide in the air sample collected from the alveoli and that in exhaled air;

....

....

[1 mark]

(ii) the percentage of nitrogen in atmospheric air and that in exhaled air.

....

....

[2 marks]

(b) (i) Describe one way in which the microscopic structure of lung tissue from a patient with emphysema would differ from that of healthy lung tissue.

....

[1 mark]

(ii) Explain how you would expect the volume of oxygen in an exhaled air sample collected from a patient with emphysema to differ from a similar sample collected from a healthy person.

....

....

[2 marks]
[Total 6 marks]

Examiner's hints

- Fick's law simply relates the three main factors that affect the rate of diffusion. You need to be able to describe how the lung is designed to have a large surface area, to maintain a diffusion gradient and to have a thin diffusion pathway.
- There are few questions that examiners can ask on this topic, and the same ideas occur again and again in different forms. Learn the principles behind diffusion and how the structure is modified for effective gas exchange.

Q2

Fick's law can be stated as:

$$\text{diffusion rate is proportional to } \frac{\text{surface area} \times \text{difference in concentration}}{\text{thickness of exchange surface}}$$

(a) With reference to gas exchange in the human lung:

(i) describe **two** features that ensure a large surface area for gas exchange;

1 ....

2 ....

[2 marks]

(ii) give **two** processes which ensure that a difference in concentration is maintained.

1 ....

2 ....

[2 marks]

(b) The diagram shows the use of the 'mouth to mouth' method of resuscitation (ventilation).

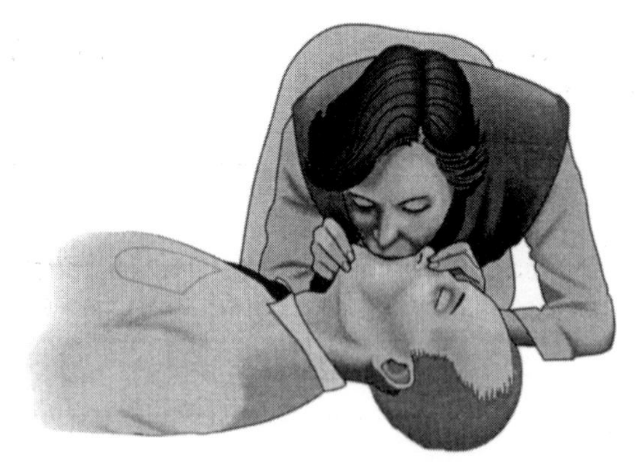

Suggest how it is possible for exhaled air to be effective when this method is used.

....

....

[2 marks]

[Total 6 marks]

Examiner's hints

- If diagrams are given that are sections through tissues or organs, try to visualise the whole thing and see how this view fits in with something you recognise.
- Unless a structure is perfectly symmetrical and the section is cut exactly through the middle you will see a whole range of different shapes depending on the plane of the cut. Imagine cutting through a whole group of identical balls trapped in a box – you would end up with circles of varying size.

The drawing shows a section through part of a human lung as it would be seen with an electron microscope.

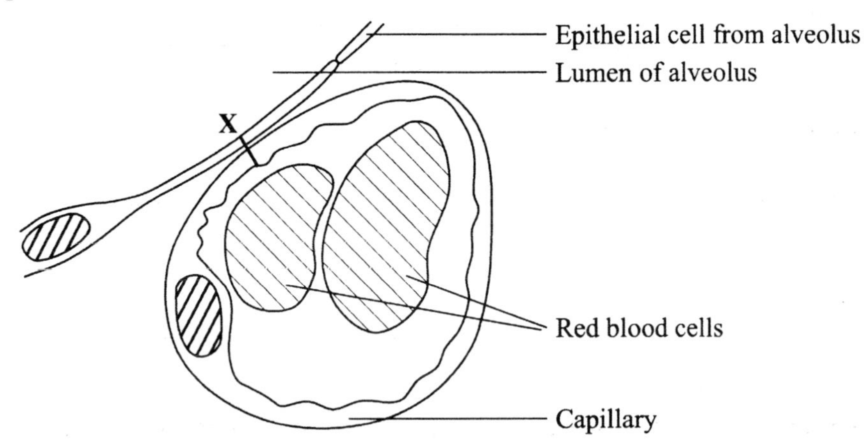

(a) (i) This drawing has been magnified 2000 times. Calculate the minimum distance that the oxygen molecules diffuse from the alveolus into the blood, shown by the line marked **X** in the diagram. Give your answer in micrometres (μm).

.... μm

[2 marks]

(ii) A thin exchange surface is one adaptation of the lungs for efficient gas exchange. Describe **one** other adaptation.

....

[1 mark]

(b) (i) In order to prepare this material for examination with an electron microscope, thin sections were cut. Explain how this results in the different shapes of the red blood cells in the drawing.

....

[1 mark]

(ii) Red blood cells are very small. Explain how this helps to ensure efficient diffusion of oxygen into the cell.

....

....

[2 marks]

Answers to Questions to try are on pp. 95 – 96.

[Total 6 marks]

10 Answers to Questions to try

Notes: A tick against part of an answer indicates where a mark would be awarded.

Chapter 1 Enzymes and their properties

Q1 How to score full marks

(a) The amount of energy needed ✓ to start the reaction.

(b) **(i)** The enzyme has an active ✓ site which has a very specific shape like the lock in a door. The shape is complementary ✓ to the shape of ✓ the substrate, which fits just as a key fits into a lock, forming an enzyme–substrate ✓ complex. This now breaks down to form the products.

Examiner's comment

The parts of the system which represents the lock and the key are identified. The importance of the matching shapes – not the same shape – is made, and the reaction is described in terms of this link. The term 'enzyme–substrate complex' is needed for maximum marks.

(ii) The induced fit model suggests that the active site is not the exact shape to fit the ✓ substrate but the contact between the two molecules causes the enzyme to change shape to fit it.

Examiner's comment

When you are comparing two processes it is vital to define your terms so you can identify which one you are describing.

(c) Oxygen is produced by the reaction and as it is a gas ✓ it escapes, reducing the mass of the beaker.

Examiner's comment

This information is given in the equation, but it is very easy not to notice it!! It is also easy to miss the point that losing oxygen will reduce the mass.

(d) As the reaction took place some of the substrate ✓ was used up. ✓ There was therefore less substrate present and less chance of collisions occurring between substrate and enzyme.

Examiner's comment

Note that as there are two marks available both 'what is happening' (**describe**) and 'why it is happening' (**explain**) answers are needed for full credit, although the question only says 'explain'.

Q2 How to score full marks

(a) **(i)** The number ✓ of substrate molecules present has ✓ decreased, and so the number of collisions between substrate and enzyme that lead to formation of the product must also decrease.

Examiner's comment

As time goes on the substrate concentration must decrease. Explain the effect this has on the turnover rate using your knowledge of how enzymes work.

(ii) It would double. At a higher temperature the molecules involved in the reaction (substrate and enzyme) would have greater kinetic energy, leading to more collisions.

Examiner's comment

Even if you say the rate will increase you will score a mark. Remember, if the question asks for an explanation, you have to write why.

(b) To show that the enzyme was responsible for the conversion.

Examiner's comment

The concept of a control is very important. In many biological systems there could be a number of **independent variables** that affect the **dependent variable** you are studying. Be sure that you control (keep constant) all the other variables and alter just one at a time.

Q3 How to score full marks

(a) **(i)** Lowering the substrate concentration would lead to fewer collisions between this and enzyme 1, so less B will be produced, and thus less D is produced.

Examiner's comment

Use the collision theory to explain how reduced amount of substrate affects the rate of a reaction. Remember that the same idea can also be used for enzyme concentration.

(ii) An excess of B would be produced, which enzymes 2 and 3 could not metabolise. Enzyme concentration will be the limiting factor and the rate of production of D will be the same.

Examiner's comment

Despite the increase in concentration of enzymes, other factors will affect the amount of end product D produced – for example, concentration of substrate A. However, if the concentration of enzymes 2 and 3 was not limiting the rate would increase. Reasoned answers along these lines would also get full credit.

(iii) There would be an increase in kinetic energy during each stage, leading to more collisions between enzyme and substrate – so more D would be produced.

Examiner's comment

Again, the links increased temperature → increased kinetic energy → increased collisions → increased rate are needed.

(b) If D binds with enzyme 1 and changes the shape or blocks the active site then it inhibits the action of the enzyme and slows the production of the end-product D.

Examiner's comment

End-point inhibition can be due to either a competitive or a non-competitive relationship between the final product of a series of reactions and the initial enzyme. This answer has described both without using these terms. It does, however, explain the effect and so is worth full marks. Sometimes, to write what is happening gives as clear an indication that you understand the biology as simply recalling a term.

Q1 How to score full marks

(a) 7.5 nm ✓

Examiner's comments
Be sure to use the right units.

(b) **(i)** Hydrophilic ✓ head ✓ of the phospholipid.

Examiner's comments
Initially it was thought that the outer layers were protein but it is now clear that the whole membrane is made of a phospholipid bilayer with protein embedded in it. The outer darker layers are the compact, denser glycerol/phosphate head of the molecule.

(ii) Hydrophobic ✓ tail ✓ of the phospholipid.

Examiner's comments
The centre of the membrane is made of two layers of fatty acid chains, which are the hydrocarbon chains of the phospholipid molecules. These molecules are less compact and therefore appear lighter.

(c) Passive ✓ movement of molecules with a ✓ concentration gradient but with the help of a carrier protein.

Examiner's comments
Facilitated diffusion requires no energy from the cell and is therefore known as passive movement. It relies on the kinetic energy of the molecules coming into contact with the membrane. The carrier protein speeds up the movement through the membrane.

Q2 How to score full marks

(a) Oxygen and carbon dioxide. ✓

Examiner's comments
Urea and other lipid soluble molecules pass through in this way too, but the respiratory gases are the easiest to remember. Both examples must be correct for the one mark.

(b) The hydrophobic layer of the membrane stops the movement of charged ✓ molecules. Sodium and potassium ions are charged atoms but glucose is not.

Examiner's comments
Most ions link with water and, although you may think that they are small, they tend to be both large and charged. Neither property allows them to move freely through the hydrophobic fatty acid layer of the membrane.

(c) Protein carriers ✓ bind with certain molecules that have complementary ✓ shapes.

Examiner's comments
There are a number of proteins, each with its own receptor site, able to join with and carry just one atom or molecule across the membrane. Never refer to these as 'active' sites.

(d) To move molecules from where there are a lot of that type molecule to where there are few. ✓

> **Examiner's comments**
>
> Remember, diffusion is the movement of molecules with a concentration gradient. Active transport is the movement of molecules against a concentration gradient.

(e) (i)

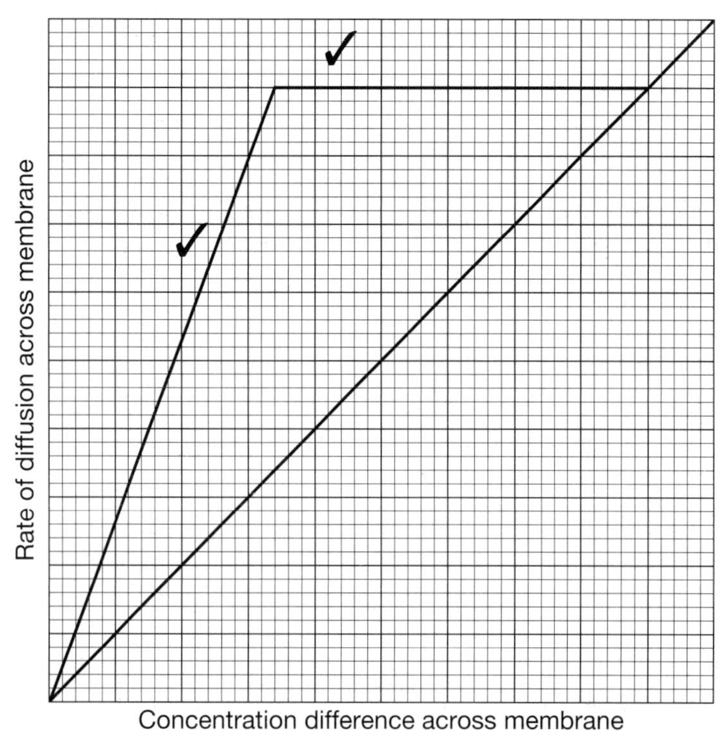

> **Examiner's comments**
>
> Facilitated diffusion causes more rapid movement of molecules through a membrane than simple diffusion – all other factors being constant – but, as it relies on protein carriers, the rate is limited by the number of proteins involved. The curve will therefore not continue to rise indefinitely.

(ii) Surface area, ✓ thickness ✓ of the membrane, ✓ temperature ✓ of the environment and the size of the molecule diffusing.

> **Examiner's comments**
>
> The table below explains why each of the features affects the rate of diffusion

Feature	Increase or decrease the rate of diffusion	Reason
Larger surface area	Increase	More particles can come into contact with the membrane and therefore pass through it during the same time period
Greater concentration gradient	Increase	The greater the difference in concentration of particles the greater the energy difference on either side of the membrane
Increased thickness of the membrane	Decrease	The thicker the membrane the greater the distance that needs to be covered to cross the membrane, and the longer it takes
Higher temperature	Increase	The higher the temperature the greater the kinetic energy each particle will have and the faster it will travel
Larger size of diffusing particle	Decrease	The bigger the particle the more resistance the membrane will offer to its movement

Chapter 3 Mitosis, meiosis and the cell cycle

Q1 How to score full marks

(a)

Mitosis	Meiosis	Neither
		✓
✓		
✓		
	✓	

Two marks for four correct ticks.

Examiner's comments

Line 1: Mitosis is the division of the nucleus so this cannot take place in bacteria.

Line 2: Even though gametes are being formed, the genetic content is not changing – a haploid cell is producing a haploid cell, which can occur only during mitosis.

Line 3: Growth needs cells with the same genetic material as the original, so mitosis is the only possibility here.

Line 4: Although the life cycle of a protoctista may not be on any syllabus, the phrase 'genetically different' is the clue here. To produce genetically different cells meiosis is needed.

(b) ✓ ✓

Examiner's comments

The pairs of chromatids moving to the top of the cell will each separate. Thus two of the four cells must receive a white chromosome (remember when pairs of chromatids separate, they can now be called a chromosome) and either a black one or a black one with a white top. The same reasoning will hold for the bottom pairs of chromatids.

(c) 8 ✓

Examiner's comments

The number of chromosomes present at the beginning of meiosis (or mitosis) is the number found in a normal body cell. Chromosomes can exist as pairs of chromatids – and do so before meiosis (or mitosis) occurs. So if 16 chromatids are present this must mean 8 chromosomes, which is the number in a normal body cell.

Q2 How to score full marks

(a) 'Anaphase' shaded ✓

Examiner's comments

Spindle fibres shorten to pull the chromatids apart, and this happens during anaphase.

(b) **(i)** During the G_1 phase protein ✓ synthesis is taking place and so there is an increase in the amount of cytoplasm present but not in the amount of material in the nucleus.

Examiner's comments

If you are asked to consider ratios write them down and see what happens if you change the figures. In this case compare the nuclear diameter to the cell diameter:

$\frac{1 \text{ (nucleus)}}{4 \text{ (cell)}}$ decrease the ratio $\frac{1}{8}$ so the cell must be getting bigger.

All you now need to do is to explain why this is happening – give the reason why.

(ii) The centrioles divide ✓ so that spindle fibres can be made during mitosis.

Examiner's comments

During G_2 division of the centrioles occurs and proteins that are involved in the formation of the spindle fibres are synthesised. Either of these answers would be acceptable.

(c) Chromosomes are attached to the spindle ✓ fibres at the equator ✓ of the cell during metaphase. During anaphase they move ✓ apart, toward the poles of the cell.

Examiner's comments

Either the names of the stages or what is happening to the chromosomes is being tested. This is a test of recall – you simply must learn it!

Q3 How to score full marks

(a) Mitosis occurs at the tip of the root because it is from here that they grow. ✓

Examiner's comments

In animal tissue mitosis occurs in all organs, but in young soft plants – organisms more acceptable to use for this sort of experiment – it occurs mainly at the tip of the root or the tip of the stem.

(b) **(i)** To allow chromosomes to be seen they must be made to contrast ✓ with the surrounding areas.

Examiner's comments

The content's of cells, including the chromosomes, are transparent. Stains are dyes that are taken up by different structures by different amounts and therefore allow those structures to be distinguished from one another. You should know that the stain used to show chromosomes is acetic orcein.

(ii) This produces a single ✓ layer of cells so that detail of the cell structure can be seen clearly.

Examiner's comments

Detail of the cellular structure of multicellular organs like the roots of plants is difficult to see if a number of cells overlie one another. Single layers are produced by cutting thin sections or by squashing.

(c) Anaphase ✓

Examiner's comments

The characteristic 'V' shape of chromatids (now called chromosomes) moving apart is seen here even though the spindle has not been shown on the diagram. Remember the word 'apart' is linked with anaphase. See the table in the key facts section on page 26.

Chapter 4 Nucleic acids

Q1 How to score full marks

(a) P – Deoxyribose sugar ✓

Q – Phosphate ✓

Examiner's comments

There are only three components of nucleic acids – and you must remember them.

(b) G with C
A with T
C with G } ✓

Examiner's comments

'A tea for two' is a useful phrase to help you recall the base pairing. **A**denine joins with **T**hymine by **two** hydrogen bonds.

(c) Thymine ✓

Examiner's comments

Thymine is replaced by uracil in all forms of RNA.

(d) Hydrogen bonds ✓

Examiner's comments

Two bonds are formed between A and T but there are three between C and G. This is the reason why adenine always pairs with thymine and not cytosine.

(e) Hydrogen bonds break ✓ and the two strands unwind. Nucleotides containing complementary ✓ bases ✓ align and join with the exposed bases of both strands. The enzyme DNA polymerase joins the nucleotides to form new complementary strands.

Examiner's comments

Base pairing occurs between **complementary** bases. This word will always get you a mark.

(f) **(i)** and **(iii)** ✓

Examiner's comments

A and T pair and therefore must occur in equal amounts – as do C and G. Therefore any equation of this form, with A on one side and T on the other, C on one side and G on the other, will be correct.

Be aware that other forms of this relationship, such as A + G/T + C = 1 represent the same idea.

But remember that the absolute values of A + T and C + G will differ with each DNA molecule because of the different base sequences.

Q2 How to score full marks

(a) A – Cytosine ✓

B – Thymine ✓

C – Adenine ✓

D – Guanine ✓

Examiner's comments

Purines and pyrimidines make up complementary base pairs. Purines are twice as large as pyrimidines. So A and B are pyrimidines, C and D are purines.

Uracil is the key here. You should remember that thymine in DNA is replaced in RNA (molecule Y) by uracil and that they are both pyrimidines. So B must be thymine. The complementary purine to thymine is adenine and the shape at the end of diagram C gives that away.

(b) **(i)** Nucleotide ✓

Examiner's comments

All nucleic acids are made up from monomers called nucleotides. DNA and RNA have different nucleotides as one of the components – the pentose sugar is different.

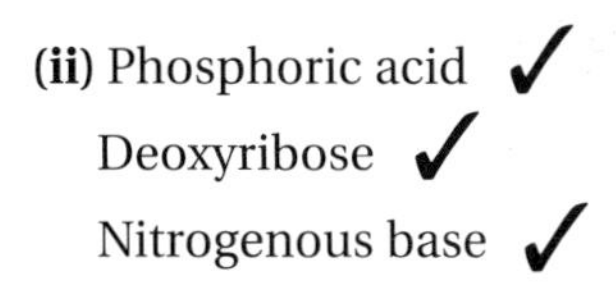

(ii) Phosphoric acid ✓

Deoxyribose ✓

Nitrogenous base ✓

Examiner's comments

If you are given terms in the stem of the question, use them. Phosphate is acceptable as an alternative to phosphoric acid. However, although deoxyribose is a pentose sugar, this is a term that also includes ribose, the sugar used to make RNA, and is thus less exact. Such a vague answer might not get you the mark. You must use the term nitrogenous base to distinguish the final component of DNA from other bases.

(c) Line drawn between any of the two bases

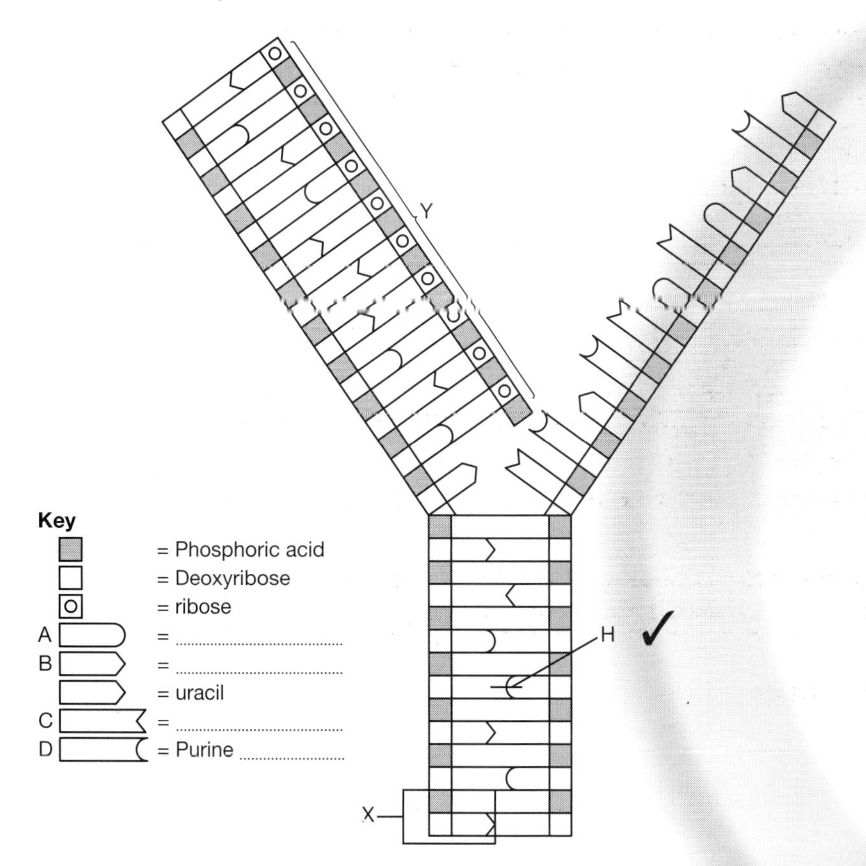

(d) Phosphoric acid and deoxyribose

Deoxyribose and nitrogenous base

> **Examiner's comments**
>
> Lots of molecules are joined by condensation reactions but this question is about nucleic acids so 'fatty acid and glycerol' or 'monosaccharide and monosaccharide' would not get you any marks.

(e) **(i)** It contains ribose not deoxyribose. 

(ii) It contains uracil not thymine.

> **Examiner's comments**
>
> You should remember these differences; questions asking for them are very popular.

Q1 How to score full marks

(a) A gene is a section of DNA and consists of a sequence of bases. These bases are decoded in sets of three – triplets. Each triplet codes for an amino acid. The sequence of the triplets therefore determines the sequence of amino acids and thus the structure of the protein.

Examiner's comments

The gene has to be related to DNA initially and then the nature of DNA – the sequence of bases, read in triplets – can be added.

(b) The sense and anti-sense strands of DNA are unwound as the hydrogen bonds linking them together are broken. The enzyme RNA polymerase assembles mRNA from nucleotides by complementary base pairing, using the sense strand of the DNA as a template. The mRNA then leaves the nucleus and associates with a ribosome. tRNA molecules carry amino acids, the particular amino acid being carried depending on the sequence of bases in the anticodon on the tRNA molecule. The anticodon on the tRNA molecule links with the complementary codon on the mRNA molecule, bringing a new amino acid into the chain. Peptide bonds form between the amino acids as they are added, and a protein is made.

Examiner's comments

Tell the whole story, following each stage in turn. There are no marks awarded for the mention of transcription and translation in this question – the examiner expects you to 'describe how' not simply to give the names of the processes. Use the correct terms – e.g. 'complementary bases', 'nucleotides', 'codon-anticodon' and 'peptide bond'. There are more marking points than those illustrated, up to the maximum of 7.

Q2 How to score full marks

(a)

G	GATCC
CCTAG	G

Examiner's comments

When hydrolysis of the bonds between nucleotides of each strand takes place the hydrogen bonds alone, in this region, are too weak to hold the strands together.

(b) The chains are held together by hydrogen bonds – two between A and T; three between C and G.

Examiner's comments

There are two marks for this section. Do not be tempted to stop after the first sentence, that alone can not be worth both marks.

(c) Using the same restriction enzyme will produce complementary sticky ends, which will allow DNA from different sources to be held together before DNA ligase joins the sugar/base backbones by condensation.

Examiner's comments

Use the correct technical terms – *complementary* ends, not 'the same' or even 'matching'. Adding the fact that the bond is formed by condensation could be an extra marking point.

Q3 How to score full marks

(a) CCT/CTG/AGG/GGG/TCA/CAT ✓

Examiner's comments

You need to get every one right to get the mark. Cover up the whole row with a ruler and expose one triplet at a time. Take your time and check at the end.

(b) glycine/asparagine/serine/proline/serine/valine ✓

Examiner's comments

If you did make a mistake in **(a)** the examiner will not 'compound your error'. If an answer to one part of a question relies on the correct answer to the previous section, any error in the first part will lead to no mark in the second. This is not fair, and questions are generally written to avoid that. If this is not possible, examiners will take the answer you give and decide if you have correctly worked out the second part of the question, but using your first wrong answer. That is why the mRNA code, the base sequence you need, is given in the stem of the question.

(c) **(i)** A sequence of three ✓ bases on mRNA

Examiner's comments

It is easy to be vague here – reference to the triplet code or of the codon consisting of three bases is vital for credit.

(ii) As a start/stop ✓ code to begin or end reading the genetic code.

Examiner's comments

The nonsense codes are vital and although none were given here they can form the basis of questions. The mutation of similar codes to create a stop code can affect the length and function of a protein.

(d) mRNA links with a ✓ ribosome ✓ and presents one of its codons. An anticodon of tRNA with complementary bases associates with the codon. No join is made but the ribosome holds these two RNA molecules together. The tRNA carries a specific ✓ amino acid, depending on the sequence of bases in the anticodon. A second ✓ codon is presented and a second tRNA brings its amino acid. A peptide bond is formed between the two amino acids and a polypeptide chain starts to form. As the amino acids join, the tRNA is freed to join with another amino acid.

Examiner's comments

If you are asked to consider tRNA and ribosomes make sure you mention them both. With four marks available you could guess that there would be two marks for each. You will have to explain the role of other relevant molecules too – mRNA in this case.

Chapter 6 Biological molecules

Q1 How to score full marks

(a) **1** Lipids are much smaller ✓ than the other macromolecules.

> **Examiner's comments**
>
> This answer may seem too easy – but always look for and state the obvious first – it is a valid difference. Too many candidates lose marks by skipping to the harder ideas and forgetting the easy ones.

2 They are not polymers ✓ – i.e. they are not made of identical units joined together. They only contain an alcohol and fatty acids.

> **Examiner's comments**
>
> To be a true polymer each monomer must be the same. For example, all of the monomers that make up the polymer cellulose are beta glucose molecules. The monomers that make up proteins (amino acids) are all slightly different – but because they all have the same basic structure we consider proteins to be polymers of amino acid monomers. That is not true of lipids, which are made up of only two types of component – glycerol and fatty acids.

(b) Rub the maize grain on paper and you end up with a translucent ✓ spot.

> **Examiner's comments**
>
> Sometimes questions ask for a chemical test and therefore in this case 'dissolve in ethanol then mix with water to give a white emulsion' would be the correct answer. If you are in doubt, the ethanol test is the best lipid test to give.

(c)

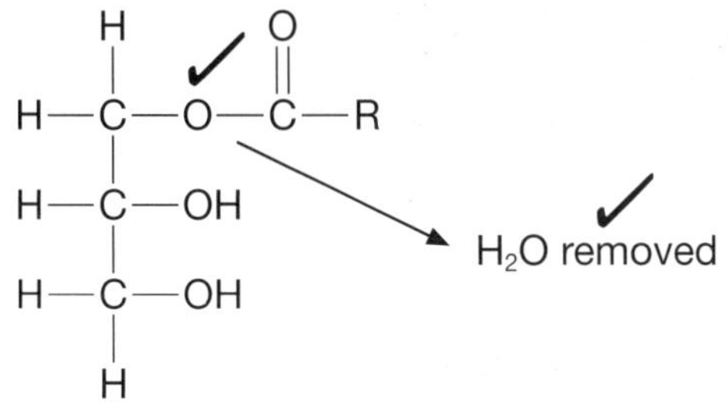

```
    H     ✓  O
    |        ||
H — C — O — C — R
    |
H — C — OH            ✓
    |          → H2O removed
H — C — OH
    |
    H
```

> **Examiner's comments**
>
> The question asks only for you to show how a monoglyceride is formed and thus attempts to save you repeating the same exercise three times. Show clearly that the bond has been formed between the OH group of the fatty acid and any of the OH groups of glycerol. The second mark in all examples like this will be for an indication that water is removed.

Q2 How to score full marks

(a) **(i)** Carbohydrates like sucrose contain the elements carbon, hydrogen and oxygen. Hydrogen and oxygen are always in a 2 to 1 ratio. ✓

> **Examiner's comments**
>
> Although the first sentence has defined the elements that make up a carbohydrate it has not distinguished it from a lipid – so that would not be enough on its own. The question gives you the clue. In most cases a skeleton diagram is given, showing only the atoms involved with bond formation. Here the OH and H groups are shown attached to each carbon atom. If you count the hydrogen and oxygen

atoms there are 22 hydrogens and 11 oxygens – so mentioning there is a 2:1 ratio will give you the mark.

(ii) The glycosidic bond is broken by boiling with acid, which produces glucose and fructose. Both these molecules are reducing sugars and will give a positive result with Benedict's reagent.

Examiner's comments

Heating sucrose with an acid will cause hydrolysis, the same process that occurs when an enzyme breaks the molecule down in the digestive system. The fact that sucrose is a non-reducing sugar is not important, only what is produced when it is broken down. A description of the test or what you would expect to see if it were positive is also not needed.

(b) There are many different proteins, each with different tertiary structures. Each different shape allows the formation of a binding site into which a different substance fits.

Examiner's comments

Proteins function as enzymes and receptors; in both cases they are able to do so because of their tertiary structure and thus their globular shape. Enzymes have **active sites** into which substrates fit – but receptors do not – they simply have **binding sites.** The term 'active site' relates specifically to enzymes.

Q3 How to score full marks

(a) (i)

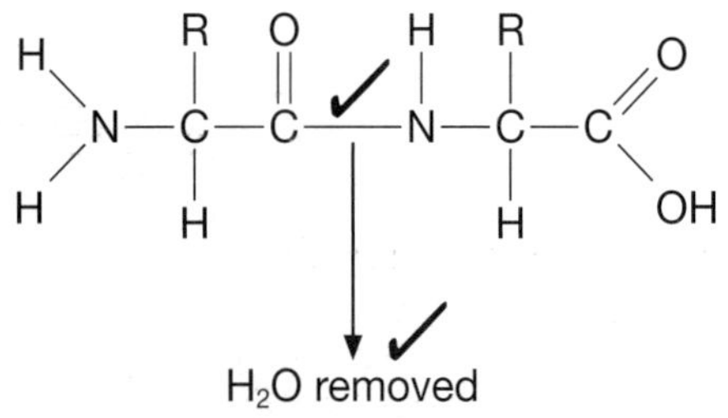

H_2O removed

Examiner's comments

The diagram must show the C–N link and that a water molecule has been removed to allow that link to form. The most common mistake is to leave the OH group intact and to link the molecules using the single oxygen atom. See the diagram below.

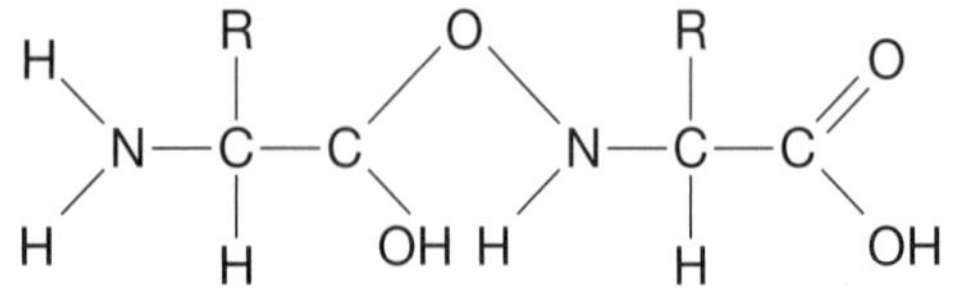

This is wrong

(ii) Condensation

Examiner's comments

The question could have been set so that the reaction name was given in the stem. It might then have asked you for the type of bond formed when two amino acids join.

(b) (i)

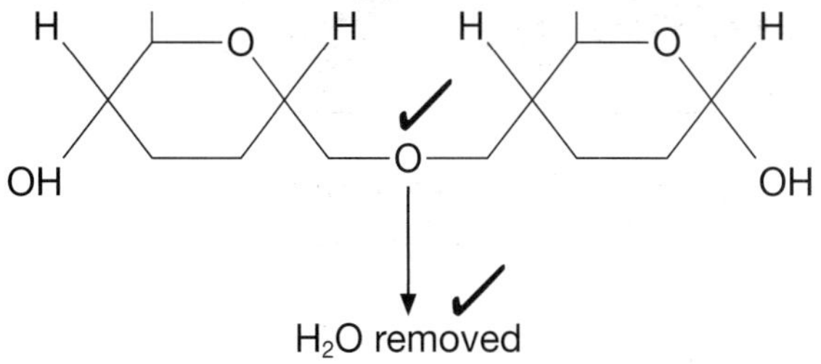

H_2O removed

> **Examiner's comments**
>
> One mark was given for the link being made between the two OH groups of the original molecules. Never join monosaccharides between the OH and H groups. The second mark was for showing that water was given off. Remember to draw the whole molecule – including the side chain and other groups shown on the original diagram. Two examples of incorrect answers, which did not get the mark, are shown below.

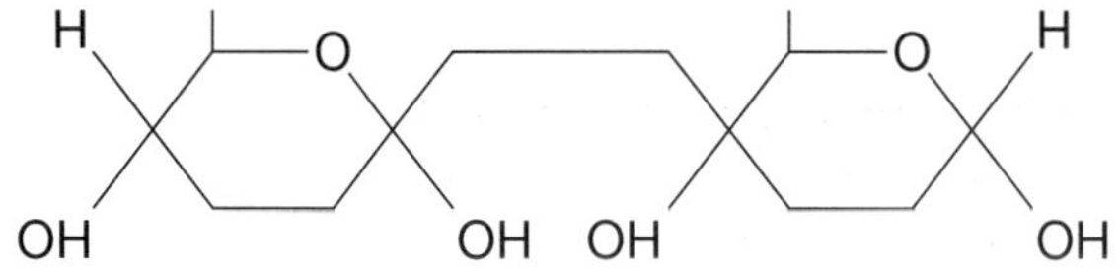

These are wrong

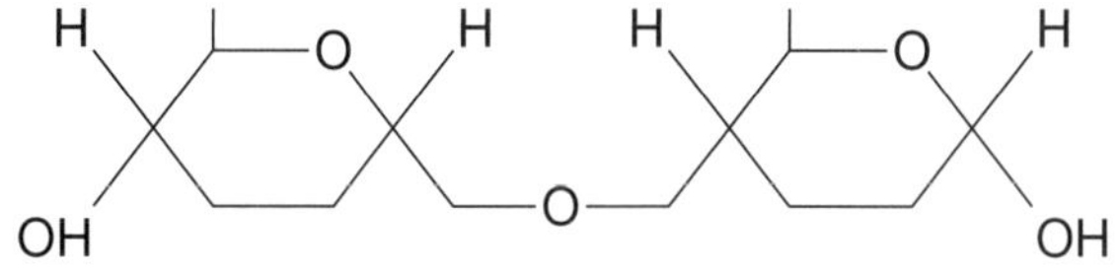

(ii) Use Benedict's reagent to test the unknown solution. If the test is positive then the solution contains glucose, if it is negative it contains sucrose. ✓

> **Examiner's comments**
>
> It was not necessary to describe the experimental details of the test, only how a test could be used to distinguish between the two carbohydrates. Another common question asks for a method of distinguishing between a carbohydrate and the enzyme which acts upon it. It is really testing two things – first whether you realise that an enzyme is a protein and secondly how you would test for the presence of a protein.

Chapter 7 Cells, organelles and sizes

Q1 How to score full marks

(a) Distance X to Y = 8 mm

Actual distance = $\frac{8}{1500}$ ✓

= 0.00533 mm

= 5.33 μm ✓

Examiner's comments

If given a magnification, (as you are here) it means that the real distance is smaller by the factor given; so divide by that factor. Do all the calculations in mm first and then convert to μm by moving the decimal point three places to the right.

(b) Cellulose is insoluble ✓ and therefore water in the environment will not dissolve it and break it down.

Examiner's comments

There are a number of different properties you could have chosen, including the fact that few organisms are able to produce enzymes to break cellulose down. Chose one you can easily relate to its function as a supporting structure.

(c) **(i)** Ribosome ✓

Examiner's comments

Ribosomes are attached to the endoplasmic reticulum in eukaryotic cells and are free in the cytoplasm of prokaryotic cells.

(ii) Golgi apparatus ✓

Examiner's comments

Proteins are transported to this organelle, where they are modified and packaged into vesicles for secretion.

Q2 How to score full marks

Fraction	Organelles	Function
A	Ribosomes	Protein synthesis ✓
B	Mitochondria	Aerobic respiration ✓
C	Nuclei ✓	Contains genetic material ✓

Examiner's comments

It is obvious that the nucleus is the densest organelle. The only problem then is to determine whether ribosomes are heavier or lighter than mitochondria. This is why only a single mark was awarded for the first column.

The more dense the organelle the further it will move down the tube. Thus the heaviest structure – the nucleus – is at the bottom and the lightest – the ribosome – is at the top.

Be aware of the function of each organelle on your syllabus – you need not be able to describe it in great detail but a description in a few words is useful.

'Respiration' alone is not good enough for the role of the mitochondria and the nucleus does not 'control cell division'.

Chapter 8 The heart and circulation

Q1 How to score full marks

(a) Right ventricle ✓

Examiner's comments

Whenever you are asked to name one of the chambers of the heart remember to give not only whether it is the atrium or ventricle but also whether it is the right or left side. There was only one mark here and 'ventricle' alone will not be enough.

(b)

Valve	Open	Closed
1		✓
2	✓	
3		✓
4	✓	

✓ ✓

Examiner's comments

At time X the pressure in the artery is rising and therefore blood must be moving through this vessel, due to the contraction of the right ventricle. If so, both cuspid valves must be open because both ventricles contract at the same time – valves 2 and 4 are therefore open. To stop blood flowing back to the atria both atrioventricular valves must be closed – valves 1 and 3 are closed.

(c) Peaks and troughs coinciding but all at a higher pressure:

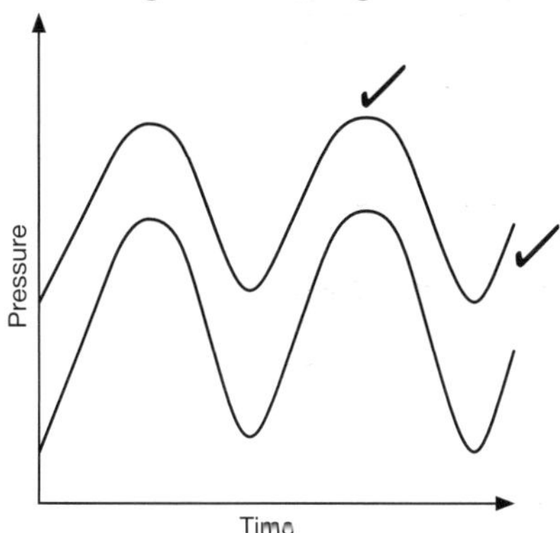

Examiner's comments

The two sides of the heart are coordinated to beat at the same time and therefore the pressure changes in both the pulmonary artery and the dorsal aorta coincide. However, the amount of muscle that makes up the left ventricle of the heart is greater than that making up the right ventricle, and creates a pressure five times higher than that produced by the right ventricle. Any line that is the same shape as the graph given but drawn above it would score both marks.

Q2 How to score full marks

(a) Right ventricle ✓

Examiner's comments

The ventricles pump blood out of the heart; the atria simply collect it from the veins and pass it to the ventricles. The left side of the heart is a high-pressure pump directing blood into arteries that carry it all over the body. The right side is a low-pressure pump directing blood the small distance into the lungs.

(b) **(i)** The ventricles are contracting and the semilunar valves are open. ✓ ✓

Examiner's comments

Pressure in this vessel is at its maximum and is about to fall. When this happens the valve will close. An answer suggesting that the valve is closing would also get credit.

(ii) The ventricles are filling with blood from the atria through the open atrioventricular valves. ✓ ✓

Examiner's comments

The pressure in this vessel is at its lowest and thus the ventricle can not be contracting. This is the other phase of the cardiac cycle, when blood is being pumped into the ventricles from the atria.

There are two marks for each of these parts, so remember to mention both which valve is open and which chamber is contracting.

(c) **(i)** The left ventricle forces blood into the dorsal aorta and then into the brachial artery, while the right ventricle pumps blood only into the pulmonary artery. The muscular walls of the left ventricle are thicker than the right, and when the left ventricle contracts it generates greater pressure in the vessel receiving the blood. ✓

Examiner's comments

Although there is only one mark here for the fact that the wall of this chamber is thicker; the term 'explain' in the question means that you will have to describe the relationship between the vessels and the chambers of the heart responsible for pumping blood into them before you can explain the difference in pressure between them. That's why such a lot of space has been left for a one-mark answer.

(ii) Blood going to the lungs through the pulmonary artery is at a lower pressure, which makes gaseous exchange more effective. ✓

Examiner's comments

The term 'suggest' means that there may be a lot of alternative answers here. 'The arms are further away from the heart than the lungs and therefore the blood needs greater pressure to travel the extra distance' is also valid.

Chapter 9 Ventilation and gas exchange

Q1 How to score full marks

(a) **(i)** Exhaled air is a mixture of air that has been in the alveoli and air from the dead space of the lung. This mixture will therefore contain a lower percentage of carbon dioxide than the sample from the alveoli.

Examiner's comments

As you exhale, you mix air from the lung, which has collected carbon dioxide from the blood, with air in the trachea, which has just been breathed in and therefore will have a low carbon dioxide percentage (the table suggests 0.04%). So the mixture will have a lower overall carbon dioxide concentration.

(ii) The relative amount of nitrogen, and therefore its percentage, will vary with the other gases in exhaled air, especially water vapour.

Examiner's comments

It is not necessary to have a moist surface through which to exchange gases. On the other hand, the gas has to be taken away from the exchange surface to maintain the concentration gradient and to allow continued diffusion to occur and this is normally achieved by dissolving it in water; water will tend to diffuse into the air in the lung. The table shows the percentage of each type of gas and therefore the relative amount of each in a given volume of the mixture. Add more of one (water vapour) without changing the absolute amount of another (nitrogen) and the percentage of the latter will fall.

(b) **(i)** Emphysema causes the breakdown of the cells making up the alveoli and therefore there are larger spaces in the lung with fewer small alveoli.

Examiner's comments

A decrease in the number of small air spaces that all blend into one another to form larger spaces will reduce the surface area available for gas exchange. The rate of diffusion is reduced by the smaller relative surface area, which produces the characteristic symptoms of breathlessness in emphysema.

(ii) There would be a greater volume of oxygen in exhaled air, as less would be absorbed due to the smaller surface area of the exchange surface of the lung.

Examiner's comments

This refers to the absolute amount of oxygen present – its volume. There would be more oxygen in the exhaled air of someone with emphysema as the rate of exchange would be less. As oxygen is exchanged for carbon dioxide there would also be less carbon dioxide in exhaled air than normal.

Q2 How to score full marks

(a) **(i)** **1** There are many small alveoli.
2 There are many tiny capillaries in contact with the alveoli.

Examiner's comments

The lung is composed of progressively smaller branched tubes, ending in air sacks. These sacks (or alveoli) are microscopic. Surrounding each alveolus are a number of fine capillaries which result in a massive air/blood surface.

(ii) **1** Ventilation of the lungs brings a fresh supply of oxygen to the lungs.
2 Oxygen dissolves in the blood and is removed by the circulation system.

Examiner's comments

It is possible to consider oxygen or carbon dioxide. In either case the principle is the same. Carbon dioxide is removed as you breathe out, keeping the concentration in the air spaces lower than that in the blood. The blood brings high concentrations of carbon dioxide as a waste product from respiring tissues, keeping the concentration in the blood at a higher level. Carbon dioxide therefore moves down the concentration gradient out of the blood and into the air in the lung.

(b) Exhaled air still contains a large volume of oxygen, at a higher concentration than found in deoxygenated blood. ✓ Thus a diffusion gradient still exists ✓ and oxygen will pass into the blood of the person being resuscitated.

Examiner's comments

The exchange of gas is not very efficient. Only about 25% of the oxygen inhaled will be taken into the blood, which means that 75% is exhaled again. There is thus still a high enough oxygen concentration to use for this method of resuscitation.

Q3 How to score full marks

(a) **(i)** Distance X is 5mm.

Actual distance is

$5 \div 2000 = 0.0025$ mm ✓

$= 2.5\ \mu m$ ✓

Examiner's comments

Measure the distance in millimetres first. Divide by the magnification to get the actual distance. To convert mm into μm (1 μm is a smaller distance than 1 mm so there will be more of them), move the decimal point three places to the right. One mark was awarded for dividing by the magnification, the other for expressing the answer correctly in micrometres.

(ii) Large surface area ✓ provided by many alveoli.

Examiner's comments

The classic answer once again.

(b) **(i)** Sections are cut through the red blood cells in different planes. ✓

Examiner's comments

The other possibility, which would also get you a mark, is that the cells are flexible and as they are squashed into the capillary they change their shape. In fact these cells, packed full with haemoglobin, are not that flexible and their movement is slowed down considerably when they are forced through the smaller capillaries. This actually helps them collect oxygen more effectively as they are in contact with the alveolar surface for a longer period.

(ii) They have a large surface area:volume ratio. ✓ Thus nowhere in the cell is very far from the surface. ✓

Examiner's comments

Red blood cells are flat biconcave discs – like thick pasta dishes with both top and bottom surfaces scooped out. This shape produces one of the highest surface area:volume ratios (the worst is a sphere) and makes rapid rates of diffusion possible.